ENERGY SCIENCE, ENGINEERING AND TECHNOLOGY

USING MUNICIPAL SOLID WASTE FOR FUEL

ENERGY SCIENCE, ENGINEERING AND TECHNOLOGY

Additional books in this series can be found on Nova's website under the Series tab.

Additional E-books in this series can be found on Nova's website under the E-books tab.

ENERGY SCIENCE, ENGINEERING AND TECHNOLOGY

USING MUNICIPAL SOLID WASTE FOR FUEL

SAMANTHA M. FELLER
EDITOR

Nova Science Publishers, Inc.
New York

Copyright © 2011 by Nova Science Publishers, Inc.

For permission to use material from this book please contact us:
Telephone 631-231-7269; Fax 631-231-8175
Web Site: http://www.novapublishers.com

NOTICE TO THE READER

The Publisher has taken reasonable care in the preparation of this book, but makes no expressed or implied warranty of any kind and assumes no responsibility for any errors or omissions. No liability is assumed for incidental or consequential damages in connection with or arising out of information contained in this book. The Publisher shall not be liable for any special, consequential, or exemplary damages resulting, in whole or in part, from the readers' use of, or reliance upon, this material. Any parts of this book based on government reports are so indicated and copyright is claimed for those parts to the extent applicable to compilations of such works.

Independent verification should be sought for any data, advice or recommendations contained in this book. In addition, no responsibility is assumed by the publisher for any injury and/or damage to persons or property arising from any methods, products, instructions, ideas or otherwise contained in this publication.

This publication is designed to provide accurate and authoritative information with regard to the subject matter covered herein. It is sold with the clear understanding that the Publisher is not engaged in rendering legal or any other professional services. If legal or any other expert assistance is required, the services of a competent person should be sought. FROM A DECLARATION OF PARTICIPANTS JOINTLY ADOPTED BY A COMMITTEE OF THE AMERICAN BAR ASSOCIATION AND A COMMITTEE OF PUBLISHERS.

Additional color graphics may be available in the e-book version of this book.

Library of Congress Cataloging-in-Publication Data

Using municipal solid waste for fuel / editor, Samantha M. Feller.
    p. cm.
  Includes bibliographical references and index.
  ISBN 978-1-61209-512-7 (hardcover)
  1. Sewage sludge fuel. I. Feller, Samantha M.
  TP360.U855 2011
  662'.87--dc23
                              2011017299

Published by Nova Science Publishers, Inc. ✛ *New York*

CONTENTS

PREFACE

Municipal solid waste (MSW) is a domestic energy resource with the potential to provide a significant amount of energy to meet U.S. liquid fuel requirements. MSW is defined as household waste, commercial solid waste, nonhazardous sludge, conditionally exempt, small quantity hazardous waste and industrial solid waste. This new book examines the potential use of MSW to make synthesis gas (syngas) suitable for production of liquid fuels; the expected process scale required for favorable economics; the availability of MSW in quantities sufficient to meet process scale requirements and state-of-the-art MSW gasification technology.

Chapter 1- Municipal solid waste (MSW) is a domestic energy resource with the potential to provide a significant amount of energy to meet US liquid fuel requirements. MSW is defined as household waste, commercial solid waste, nonhazardous sludge, conditionally exempt, small quantity hazardous waste, and industrial solid waste. It includes food waste, residential rubbish, commercial and industrial wastes, and construction and demolition debris. It has an average higher heating value (HHV) of approximately 5100 btu/lb (as arrived basis).

Chapter 2- Biomass is a renewable energy resource that can be converted into liquid fuel suitable for transportation applications and thus help meet the Energy Independence and Security Act renewable energy goals (U.S. Congress 2007). However, biomass is not always available in sufficient quantity at a price compatible with fuels production. Municipal solid waste (MSW) on the other hand is readily available in large quantities in some communities and is considered a partially renewable feedstock. Furthermore, MSW may be available for little or no cost.

Chapter 3- Heightened interest in renewable energy has prompted the Energy Information Administration (EIA) to examine some aspects of how it classifies energy sources as renewable. EIA employs the following definition of renewable energy sources: "Energy resources that are naturally replenishing but flow-limited. They are virtually inexhaustible in duration but limited in the amount of energy that is available per unit of time. Renewable energy resources include: biomass, hydro, geothermal, solar, wind, ocean thermal, wave action, and tidal action." Note that this definition defines renewable energy according to its primary source, which contrasts with other definitions that define any recurring waste stream as renewable.

In: Using Municipal Solid Waste for Fuel   ISBN: 978-1-61209-512-7
Editor: Samantha M. Feller   © 2011 Nova Science Publishers, Inc.

Chapter 1

MUNICIPAL SOLID WASTE (MSW) TO LIQUID FUELS SYNTHESIS, VOLUME 1: AVAILABILITY OF FEEDSTOCK AND TECHNOLOGY

United States Department of Energy

EXECUTIVE SUMMARY

Municipal solid waste (MSW) is a domestic energy resource with the potential to provide a significant amount of energy to meet US liquid fuel requirements. MSW is defined as household waste, commercial solid waste, nonhazardous sludge, conditionally exempt, small quantity hazardous waste, and industrial solid waste. It includes food waste, residential rubbish, commercial and industrial wastes, and construction and demolition debris. It has an average higher heating value (HHV) of approximately 5100 btu/lb (as arrived basis).

According to the United States Environmental Protection Agency (EPA), the annual national MSW production in 2006 totaled more than 251 MM short tons, which equates to greater than 4.5 lbs/person/day. Of this total, about 45% is recovered via recycling, composting, and energy production. This leaves approximately 138 MM short tons of unutilized MSW, which has about 1.4×10^{15} Btu (1.4 quadrillion Btu) fuel value associated with it.

MSW has potential as a gasifier feedstock because it has a HHV (dry basis) that is nearly as high as most conventional biomass feedstocks. What makes it appealing as a potential feedstock is that it is readily available in the near-term having a pre-existing collection/transportation infrastructure and fee provided by the supplier that does not exist for conventional biomass resources.

This report investigated the potential of using MSW to make synthesis gas (syngas) suitable for production of liquid fuels. Issues examined include:

- MSW physical and chemical properties affecting its suitability as a gasifier feedstock and for liquid fuels synthesis
- expected process scale required for favorable economics
- the availability of MSW in quantities sufficient to meet process scale requirements
- the state-of-the-art of MSW gasification technology.

MSW is a heterogeneous feedstock containing materials with widely varying sizes and shapes and composition. This can be difficult to feed into many gasifiers and can lead to variable gasification behavior if used in an as received condition. It is expected that some minimal size reduction and sorting will need to be performed to make MSW suitable as a feedstock for MSW gasifiers. Refuse derived fuel (RDF) is a processed form of MSW where significant size reduction, screening, sorting, and, in some cases, pelletization is performed to improve the handling characteristics and composition of the material to be fed to a gasifier. The trade-off between the increased costs of producing RDF from MSW and potential cost reductions in gasifier design and operation are explored.

The chemical make-up of MSW includes significant quantities of chemical constituents that can create problems in downstream processes. While the concentrations of these contaminants are greater than that found in conventional biomass feedstocks, they are roughly comparable to those found in coal.

In order for MSW to be used for liquid fuels synthesis, it must be available in sufficient quantities to meet the necessary process scale for favorable economics. Using a previous economic study on liquid fuels synthesis from gasification of wood, the expected process scale could by as much as 3,300 short tons per day (as arrived basis). However, the economics associated with MSW are very different from conventional biomass feedstocks. Whereas conventional biomass costs about $44/dry ton to purchase as a feedstock, landfills around the country using incinerators and gasifiers to produce electric power from MSW charge about $30/ton (as received) additional tipping fee beyond that typically charged to just landfill the material. By charging this additional tipping fee, it may be possible to achieve favorable economics for liquid fuels synthesis at a much smaller scale.

This study identified 47 landfill sites around the country that processed 3300 short tons per day or more of MSW (as arrived basis). Together these sites could produce more than 310,000 bbl per day of liquid fuels (as fuel grade ethanol), which is equal to about 1.4% of current U.S. transportation fuels demand. This would double if plants as small as 1250 short tons per day were found to be economical with appropriate tipping fees. This study (and others) identified that United States MSW landfills receive more than double the mass of material estimated using values reported by the EPA. Reasons for this discrepancy are discussed.

This study also identified four commercially available and two demonstration MSW gasifiers, two of which involved co-feeding with coal. Three of these gasifiers were used to produce electricity and process heat and consequently were close coupled to combustors and would need to be modified to produce syngas. The other three gasifiers were capable of producing syngas or fuel gas, and would not require modification. There were also a large number of other MSW gasifiers under development that may ultimately be suitable for syngas applications.

Overall, this study concludes that MSW should be considered as a potentially viable gasifier feedstock for liquid fuels synthesis. A review of feedstock availability, composition, and handling characteristics along with commercially available MSW specific gasifiers did not identify any obvious insurmountable technical or economic barriers to commercialization. However, further research into the economic issues surrounding tipping fees and process scale is needed to verify economic viability and the appropriate plant scale to achieve it.

ACKNOWLEDGMENTS

This report is a survey of the information available on municipal solid waste and biomass gasification as a pathway for synthesis of liquid fuel. The authors did extensive research to locate the information as listed in the Reference section of this document. Sources for all data shown in figures and tables are referenced within the text. Tables and figures not generated by the authors come from their identified source.

ACRONYMS AND ABBREVIATIONS

EPA	U.S. Environmental Protection Agency
Ecology	Washington State Department of Ecology
MSW	Municipal Solid Waste
Btu	British thermal unit
MJ	Megajoule
MM	Million
kg	Kilogram
RDF	Refuse Derived Fuel
bbl	Barrels
LFG	Landfill Gas

1.0. INTRODUCTION

The United States produces the largest amount of municipal solid waste (MSW) per capita in the world. The United States Environmental Protection Agency (EPA) estimated that the national MSW production totaled more than 251 MM short tons in 2006, before recycling, which equates to greater than 4.5 lbs/person/day. For perspective, assume an average heating value of 5100 Btu per pound of MSW. Approximately 2.5×10^{15} Btu (2.5 quadrillion Btu) per year would then be associated with the U.S. MSW stream. Energy Information Administration (EIA) estimates for total U.S. energy consumption, broken into the major sectors for 1980 to 2008 is shown in Figure 1.1 for comparison (DOE/EIA 2008). The heating value of the EPA's estimated national MSW stream is equal to 21%, 29%, 8%, and 10% of the EIA's estimates for residential, commercial, industrial, and transportation sector's energy consumption, respectively, or more than 3% of EIA's estimate for total U.S. energy demand.

Energy Consumption by Sector 1980-2008 (Quadrillion Btu)

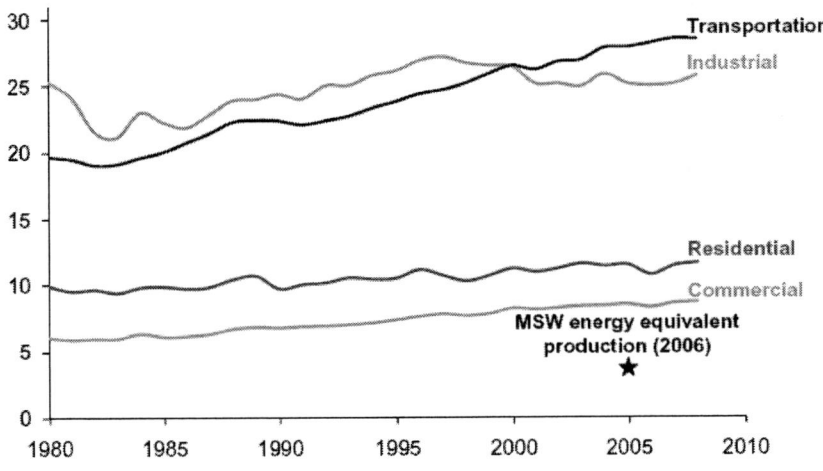

Figure 1.1. Energy Consumption by Sector, 1980-2008.

Not all of this material may be considered to be available. The EPA reported that recycling (including composting) and waste-to-energy facilities diverted 113 MM short tons from landfill disposal in 2006. Therefore, a more realistic approximation may be that 55% of the U.S. MSW stream is available for conversion. Using the same assumptions for heating value and EIA estimates for energy demand, the heating value of the theoretically available MSW stream would be equivalent to 11%, 16%, 4%, and 5% of the nation's residential, commercial, industrial, and transportation sectors, respectively. This interpretation is a straight comparison of the heating value associated with the EPA's estimates for landfilled MSW with energy demand and does not include any loss in efficiency associated with conversion to any particular fuel (e.g., power, liquids).

While the estimated energy content of MSW currently entering landfills is a small fraction of the total amount of energy consumed by the U.S., recent studies show that the organic fraction of MSW represents a significant resource when compared to biomass fuels currently available for conversion to energy. For example, MSW accounts for 70% of the carbonaceous resource (including MSW and biomass) currently available in areas such as Oregon and Washington (Stiles et al. 2008). Since the major fraction of MSW consists of biomass materials, utilization of MSW also provides environmental benefits, such as reduction of green house gas emissions (CO_2, CH_4) and landfill space, as well as reduced environmental hazards potential associated with soil and water contamination and unplanned fires at landfills.

Conventional incineration (and gasification) technologies, often referred to as "waste-to-energy" technologies, are currently used in some states to generate power and heat. Gasification technologies enable conversion of MSW into value added products, such as liquid fuels and commodity chemicals, as well as electricity, and do so at greater efficiencies than conventional incineration (Stevens 1994, Klass 1998, Rezaiyan and Cherenmisinoff 2005). This report considers the nation's MSW as a potential carbonaceous feedstock for conversion to fuels and chemicals. An initial assumption is that the MSW infrastructure (e.g., collection, transportation), already in place, makes landfills ideal locations for near- term alternative fuel synthesis technologies. The latest developments in gasification that are key to alternative fuel synthesis are also reviewed.

1.1. Municipal Solid Waste

Municipal solid waste is defined as household waste, commercial solid waste, nonhazardous sludge, conditionally exempt, small quantity hazardous waste[1], and industrial solid waste (EPA-530-R-95-023 1995). MSW includes food waste, rubbish from residential areas, commercial and industrial wastes, and construction and demolition debris. Biosolids (byproducts from waste water treatment, also known as sewage sludge) are not included in the formal definition of MSW, though it is estimated that 20 percent of the 8 MM dry short tons of biosolids generated annually are handled by MSW landfills (EPA-530-R-99- 009 1999, EPA-832-R-06-005 2006).

Gross MSW composition is typically documented under the categories of "Products" and "Other Wastes". The "Products" category includes paper, glass, metal, plastic, rubber and leather, textiles, wood, and other materials generated by the consumer market. "Other Wastes" includes food, yard trimmings, and other organic wastes. Defining these particular subcategories assists reporting agencies, such as the EPA, in setting benchmarks for landfill diversions (e.g., recycling, composting, and waste-to-energy). Table 1.1 shows MSW generation rates for 2006 (EPA-530-F-07-030 2007).

**Table 1.1. Gross MSW Composition, as Presented
by the U.S. EPA for 2006**

Materials	Mass Generated (MM sTon)	Percent of Total Generation
Paper and Paperboard	85.3	33.9
Glass	13.2	5.3
Metals		
Ferrous	14.2	5.7
Aluminum	3.26	1.3
Other Nonferrous	1.65	0.7
Total Metals:	19.1	7.6
Plastics	29.5	11.7
Rubber and Leather	6.54	2.6
Textiles	11.8	4.7
Wood	13.9	5.5
Other Materials	4.55	1.8
Total Materials in Products:	**164.79**	**73.2**
Other Wastes		
Food Scraps	31.3	12.4
Yard Trimmings	32.4	12.9
Miscellaneous Inorganic Wastes	3.72	1.5
Total Other Wastes:	**67.42**	**26.8**
Total MSW:	**251.31**	**100**

The generation rates given in Table 1.1 are compiled using a materials flow methodology (Franklin Associates 2008). In other words, it is assumed that all materials produced are eventually recycled, combusted, or disposed of. Production rates are obtained from "industry associations, key businesses, and similar industry sources" (EPA 2008). Adjustments are made to account for lifetimes of products, the import/export of materials, wastes and other "diversions" (EPA 2008). The database and its corresponding model have been developed and refined over the last 40 years (Franklin Associates 2008) and are a reasonable approach to estimating national MSW generation rates. This top-down approach leads to a total of 251 MM short tons of waste generated, with 82 MM short tons recycled and 31 MM short tons used for waste-to-energy in 2006. A first approximation is that the left-over mass, 138 MM short tons, entered the nation's landfills that year.

The materials flow methodology is appropriate for establishing national MSW trends by which recycling (and other diversions) may be compared and targets established. However, this methodology does not yield enough details for determining the feasibility of using MSW for conversion to liquid fuels or chemicals, nor does it provide MSW distribution information specific to a given region. Published EPA MSW "generation rates" differ from the rates at which the entirety of waste components actually enters the waste stream infrastructure. For example, values given in Table 1.1 do not include construction and demolition (C&D) debris, industrial process wastes, or certain other wastes entering landfills through this infrastructure. C&D lumber, alone, is reported to comprise as much as 10 percent of the bone dry MSW stream in states like California and Washington (California Biomass Collaborative 2006, Ecology 2006). Non-MSW wastes entering the nation's landfills results in local and state governments reporting MSW rates that are more than double the amount published by the EPA.

A regional waste characterization study such as the one found for California (340-04-05 2004) should be used to define the gross biomass, other organic, and inorganic fractions of that waste stream. A bottom-up analysis approach is essential for capturing all of the waste stream components. Databases providing compositional data for biomass and waste materials (Phyllis 2008) are available to aid the assessor in evaluating a potential conversion stream by incorporating ash, moisture, and heating value into the facility siting and economic assessments. By using these types of information, a table more amenable to assessing MSW conversion opportunities can be prepare as was the information for California shown in Table 1.2, in which quantities produced, heating values, and inorganic fractions are clearly defined for key categories of waste.

2.0. Refuse Derived Fuel

Refuse derived fuel (RDF) is defined as the product of a mixed waste processing system in which certain recyclable and non-combustible materials are removed with the remaining combustible material converted for use as a fuel to create energy (EPA-530-R-95-023 1995). Typical RDF processing includes ferrous material removal, shredding, screening, crushing, and even eddy-current separation or air classification for aluminum recovery. Some operations further grind/shred, and mix material to generate a homogeneous fuel. It is also common for processors to press and extrude the material into pellets.

On average, 75%–85% of the weight of MSW is converted into RDF and approximately 80%–90% of the BTU value is retained. This leaves RDF with a higher heating value of 4,800–6,400 Btu/lb, which is approximately half of the Btu value of the same weight of coal (NREL/TP-43 1 -4988A 1992). The main benefits of converting MSW to RDF are a higher heating value, more homogeneous physical and chemical compositions, lower pollutant emissions, reduced excess air requirement during combustion, and finally, easier storage, handling, and transportation.

The mass balance and quality of the end product are heavily affected by the type, number, and position of the equipment for these steps. Figure 2.1 illustrates a typical line configuration for a RDF production train. Shredders typically use hammers running at high speeds to chop and shred the waste. Trommel screens are rotating, tubular vessels used for sorting the waste by size. Eddy current separators use induced magnetic fields to create repelling forces that eject non-ferrous metals into a bin separate from the remaining materials. The magnetic separator uses magnets to remove ferrous metals that can ultimately be recycled. Finally, mills (again, often hammer mills) shred the material into fine particles. Additional stages can be included at various points in the system including hand sorting and air classification. Air classification, when used, can separate glass from metal free MSW.

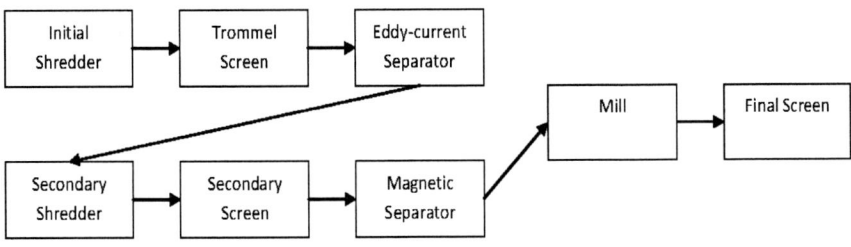

Figure 2.1. Typical RDF Production Line.

Table 1.2. Example MSW Characterization Table (California Biomass Collaborative 2006)

Material	Landfilled Mass, wet (MM sTon)	% of Total (wet) MSW	Moisture Content (wt.%)	Landfilled Mass (MM BDT)	Ash (wt.%, dry)	Ash (MM sTon)	HHV (MJ/kg dry)	HHV (MJ/kg daf.)	HHV (MJ/kg ar)
Biomass									
Paper/Cardboard	9.1	21.0	5.2	8.7	7.9	0.7	21.2	23.1	20.1
Food	6.4	14.6	36.6	4.0	5.4	0.2	-	23.7	-
Leaves and Grass	1.8	4.2	43.4	1.0	9.2	0.1	-	19.6	-
Other Organics	1.9	4.4	26.4	1.4	10.6	0.1	-	20.9	-
C&D Lumber	4.2	9.6	12.9	3.6	9.9	0.4	-	20.4	-
Prunings and Trimmings	1.0	2.3	31.0	0.7	6.8	0.0	-	19.6	-
Branches and Stumps	0.1	0.3	46.7	0.1	3.2	0.0	20.8	21.5	11.1
Total Biomass Carbon Compounds:	**24.5**	**56.4**	**28.9**	**19.5**	**7.6**	**1.6**	**21.0**	**21.2**	**15.6**
Other Organics									
All non-Film Plastic	2.2	5.1	0.2	2.2	2.3	0.1	41.9	42.9	41.8
Textiles	1.9	4.3	13.5	1.6	16.2	0.3	21.9	26.1	18.9
Film Plastic	1.9	4.4	0.2	1.9	0.1	0.0	40.6	40.7	40.6
Total Non-Biomass Carbon Compounds:	**6.0**	**13.8**	**4.6**	**5.7**	**6.2**	**0.3**	**34.8**	**36.6**	**33.8**
Inorganic									
Other C&D	5.3	12.1	-	5.3	100.0	5.3	-	-	-
Metal	3.3	7.7	-	3.3	100.0	3.3	-	-	-
Other Mixed and Mineralized	3.3	7.7	-	3.3	100.0	3.3	-	-	-
Glass	1.0	2.3	-	1.0	100.0	1.0	-	-	-
Total Mineral:	**13.0**	**29.8**	-	**13.0**		**13.0**	-	-	-
Total Landfilled MSW:	**43.5**	**100.0**	**21.6**	**38.2**	**33.7**	**14.8**	**29.3**	**25.8**	**26.5**

The process flow diagram shown in Figure 2.1 yields a low-density product fuel known as fluff. Densifiers or pelletizers are commonly added at the end of a system to generate an easily stored and transported fuel. Due to the large variance in RDF production lines, the efficiency of the process as well as the heating value of the product varies. The production line illustrated in Figure 2.1 has a mass efficiency (i.e., ratio of produced RDF mass to inlet waste mass) of approximately 24%. Note that typical mass efficiencies are reported to range from 1 8%-3 1% (Caputo and Pelagagge 2001). On average, 0.031-0.046 MM Btu per short ton of MSW is required for RDF production (NREL/TP-43 1-4988A 1992). The value would be somewhat higher when the RDF is further pelletized, creating a trade-off between improved material handling properties and energy expended to make the pelletized product.

3.0. MSW'S SUITABILITY FOR GASIFICATION INTO A SYNTHESIS GAS FOR LIQUID FUELS PRODUCTION

The suitability of MSW for gasification into a synthesis gas for liquid fuels production depends on the quality of MSW both as a feedstock for a gasifier and in terms of the gas cleanup requirements needed to removing impurities from the gas that that can cause problems in downstream processes, and the overall economics of the process. The latter is important because it generally dictates the minimum scale for the process.

3.1. MSW Feedstock Quality Issues

The quality of MSW as a feedstock to a gasification process is important in terms of gasifier design requirements to handle the feedstock and gas cleanup equipment needed to remove contaminants in the gas leaving the gasifier that can be detrimental to down-stream catalytic processes. The physical characteristics of MSW have a direct effect on gasification in terms of the degree of difficulty of feeding MSW into a gasification vessel and removing ash and other tramp inorganic material from the vessel. Conventional biomass gasifiers use biomass feedstocks that have been chipped or hammer milled into a size that can be auger-fed into the gasifier. Ash produced during gasification is either removed as fly ash from the product gas using cyclones or filters, or is removed from the bottom of the gasifier vessel using another auger. MSW contains materials ranging in size from dusts and liquids to large objects made of metal, plastic, wood and/or other materials as indicated in Tables 1.1 and 1.2. In order to use MSW as a feedstock, it either needs to be reduced in size so that it can be fed into a gasifier using an auger, or the gasifier feed system needs to be designed handle larger objects. In either case, some minimal processing is usually needed to remove very large objects made of inorganic materials. If relatively large inorganic objects are fed into the gasifier, provisions must be made in the gasifier design for removal of large pieces of inorganic materials from the bottom of the gasifier along with more conventional fine ash. Sorting processes may be used to remove very large objects and some minimal size reduction via shredding or compaction to provide a more manageable feedstock. As was previously discussed, one of the purposes of making RDF from MSW is to produce a feedstock that is more amenable to conventional biomass feeder systems. Any processing of the MSW to produce a more easily handled material will add cost to the overall gasification process, creating an economic trade-off between feed preparation costs and gasifier capital and operating costs.

The chemical composition of MSW also has an impact on its suitability as a gasifier feedstock. As previously discussed, both MSW and RDF contain a significant amount of inorganic material. Figure 3.1 shows a comparison of MSW and RDF to more conventional biomass feedstocks and coal (Phyllis 2008). It can be seen that both MSW and RDF have ash contents that are much higher that other biomass feedstocks and about 30% higher than coal. MSW also has a much greater water content than the other biomass and coal feedstocks, while RDF has a water content that is more comparable to the other feedstocks. High ash and water content in MSW feedstocks is detrimental because it dilutes its higher heating value (HHV) as shown in Figure 3.2 (HHV comparison on an as received (ar) basis) (Phyllis 2008). In addition to diluting the heating value of the feedstock, water also creates an energy burden to the overall process in order to dry the feedstock prior to gasifying it. RDF, because of its lower moisture content, has a higher as received HHV than MSW even though their ash content is nearly the same. Figure 3.2 shows that on a dry basis both MSW and RDF have HHVs that are comparable to the other forms of biomass. Interestingly, on a dry ash free basis (daf), both MSW and RDF have higher HHVs than the other forms of biomass, due to its higher content of non-biomass derived organic materials such as plastics and rubber. This is further illustrated in a graph based on New York City solid wastes, by Themelis et al. (2002), shown in Figure 3.3.

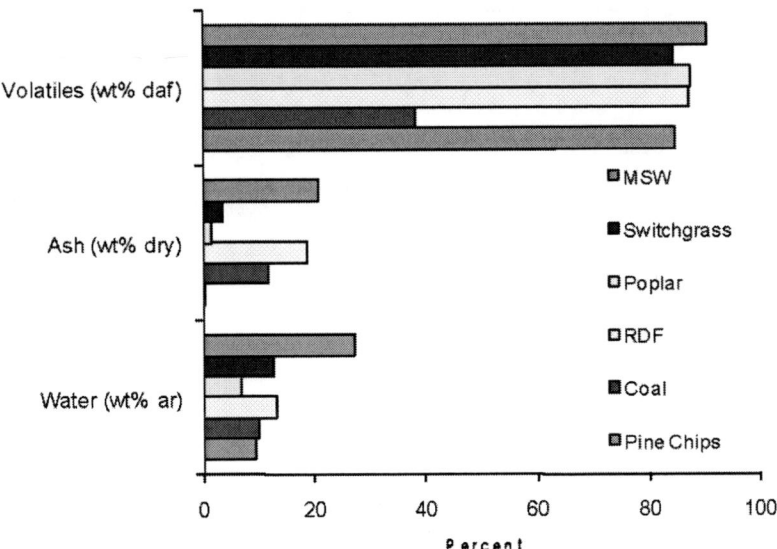

Figure 3.1. Comparison of RDF and MSW with Conventional Gasifier Fuels.

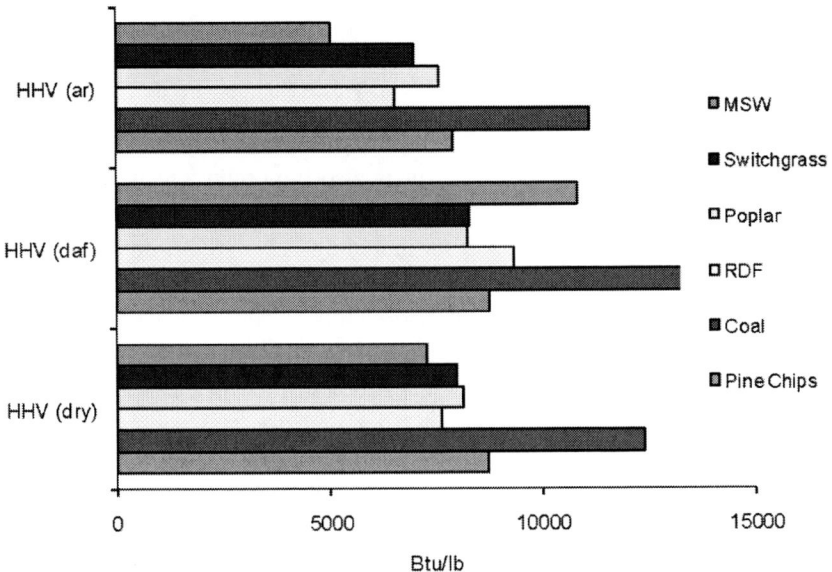

Figure 3.2. Heating Value Comparison of RDF and MSW with Conventional Gasifier Fuels.

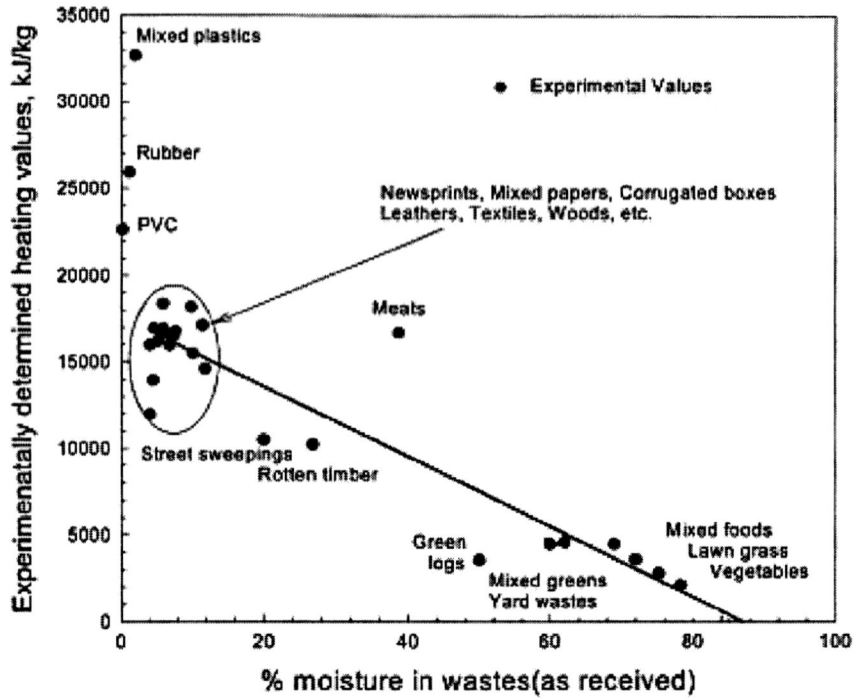

Figure 3.3. Variation of Heating Values of Solid Waste Components as a Function of Moisture Content (Themelis et al. 2002).

The chemical composition of MSW is also important because it contains many substances that can produce gas contaminants that create problems for downstream processes using catalysts including steam reforming and water-gas shift, and liquid fuels syntheses. A summary of compiled elemental analysis for bulk MSW (Phyllis 2008) is given in Table 3.1. This data provides a gross estimation of the concentrations of selected elements that can be found in MSW. Of particular note are S, Cl, F, As and P, which can be present as gaseous constituents in the product gas from a gasifier and are known catalyst poisons for liquid fuels synthesis processes. Other elements of note are Cd and Hg, which are volatile metals that may be difficult remove and are also potential catalyst poisons. The data presented in Table 3.1 has limited applicability to specific sites. It may be used as a first approximation for raw MSW, but cannot reflect streams where fractions have been separated out as a gasification feedstock, such as in the production of RDF. Figure 3.4 shows a comparison of the general elemental composition of MSW with RDF, common biomass fuels, and coal (Phyllis 2008).

Table 3.1. Elemental Composition of Bulk MSW

Component	Metric	Mean	Min	Max	# of References
Water	wt% wet	14.6	2.9	38.7	26
Volatiles	wt% daf	88.7	74.6	99.4	22
Ash	wt% dry	17	4.4	44.2	33
HHV	kJ/kg daf	24597	13130	44029	35
LHV	kJ/kg daf	22915	12126	10986	33
C	wt% daf	54.8	33.9	84.8	34
H	wt% daf	8.12	1.72	15.16	34
O	wt% daf	34	15.8	43.7	33
N	wt% daf	0.94	0.12	2.37	34
S	wt% daf	0.4	0.006	1.4	30
Cl	wt% daf	0.716	0	1.558	25
F	wt% daf	0.014	0	0.043	3
Br	wt% daf	0.001	0	0.002	3
Al	mg/kg dry	1600	1600	1600	1
As	mg/kg dry	6.9	1.5	15	5
Cd	mg/kg dry	13.6	1	35	5
Co	mg/kg dry	46.7	0.1	130	5
Cr	mg/kg dry	94.6	8	240	5
Cu	mg/kg dry	325	35	750	5
Fe	mg/kg dry	752.7	490	1000	3
Hg	mg/kg dry	0.6	0.1	2	5
Mg	mg/kg dry	120	100	130	3
Mn	mg/kg dry	156.8	10	290	4
Mo	mg/kg dry	29	2	50	3
Ni	mg/kg dry	59.6	1	150	5
P	mg/kg dry	546.7	40	850	3
Pb	mg/kg dry	226	50	350	5
Sb	mg/kg dry	13.3	10	20	3
Sn	mg/kg dry	0.1	0.1	0.2	2
Ti	mg/kg dry	145	100	190	2
V	mg/kg dry	37.3	4	75	4
Zn	mg/kg dry	306.3	85	500	4

1 kJ/kg = 0.43021 Btu/lb
1 mg/kg = 1 ppm

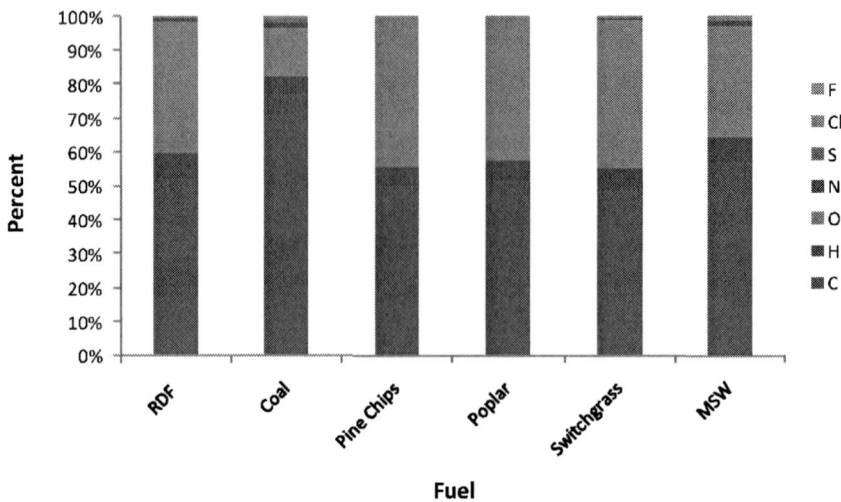

Figure 3.4. Elemental Comparison of RDF and MSW with Conventional Gasifier Fuels.

3.2. Economic Issues Affecting MSW Suitability as a Feedstock

Another important issue affecting the suitability of MSW as a feedstock for liquid fuels synthesis is its cost and how that cost affects the necessary plant scale needed to economically produce the liquid fuels. For example, a 2008 study showed that a conventional biomass (poplar wood) to mixed alcohol synthesis facility becomes economically feasible at a scale of 2,000 dry tonnes (2200 short tons) per day (Stiles et al. 2008). Using the HHV of biomass feedstock as the basis for estimating the product output from a liquid fuels synthesis facility, a rough estimate can be made regarding the amount of MSW or RDF feed that would be required for a similarly sized facility based on product output. Using the HHVs (dry basis) in Figure 3.2, a 2,200 dry short ton per day biomass fed facility would require approximately 2,400 dry short tons per day of dry MSW or 2,300 dry short tons per day of dry RDF to supply the same quantity of stored energy to the gasifier. Using the water content values in Figure 3.1, the as arrived MSW and RDF feedstock requirements would be approximately 3,300 and 2,755 short tons per day, respectively. MSW required to supply 2,755 short tons per day of RDF would range from 3,200 to 3,400 short tons per day based on previous discussions on MSW to RDF conversion.

The actual plant scale for an MSW fed process may not be the same as that for a biomass fed process because the cost of the feedstock must be taken into account. The biomass study discussed above assumed a $48.46/dry tonne ($44/dry short ton) cost and accounts for a significant fraction of the total product cost. If the biomass cost could be significantly reduced then the reduced operating costs could off-set the diseconomies of scale of a smaller plant so that a smaller scale plant could still produce a product with the same selling price.

Biomass conversion facilities using traditional forms of biomass feedstocks, such as wood chips, must pay a market price based on the value of the biomass. This value considers the cost to produce and/or collect the biomass and its open market value as feedstock. Consequently, a plant operator has limited flexibility in the price paid for biomass. MSW disposal and conversion facilities, however, charge a tipping fee for MSW based on the cost to dispose of the waste. The tipping fees charged for MSW are regulated, at least in part by local governing agencies such as cities or counties, mostly because many local municipalities also own their own landfills. As such, they may allow a reasonable profit after paying for the necessities; sorting and recovery of recyclables (if performed), incineration of a portion of the recovered materials (if performed), any unrecovered material, incinerated ash disposal, and future site remediation. Unless a flow control is in place, which would require full support of the local municipality, this scenario would promote reasonable tipping fees based on the recovery/disposal pathway costs by surrounding landfill management entities.

Tipping fees vary widely across the country. A tipping fee survey by (Repa 2005) showed that regional average tipping fees in 2004 ranged from a low of $24.06 per ton in the south to $70.06 per ton in the Northeast. Tipping fees within a state can also vary widely. For example, in 2000, county charges within Florida ranged from $23.00 per ton in Manetee County to $92 per ton in Monroe County (Florida Department of Environmental Protection 1999). Lower tipping fees are thought to reflect continued use of older, established landfills built prior to Resource Conservation and Recovery Act (RCRA), Subtitle D (40 CFR Part 257, 40 CFR Part 258)regulations that were put into effect October 9, 1993. RCRA regulations are also thought to render small landfills (<500 short tons per day) prohibitively expensive to build and operate (EPA-530-R-99-016 1999, EPA-530-R-99-013 1999).

Understanding regional flows, with such a wide range in tipping fees structure, is not straightforward. This is because state and local governments began implementing flow controls, in the late 1970s, to spur the development of new MSW capacity. Flow controls are defined by the EPA as "legal authorities used by State and local governments to designate where MSW must be taken for processing, treatment or disposal" (EPA-530-R-95-008 1995). A direct effect of such flow controls is assurance that designated facilities receive a guaranteed amount of MSW (and/or recyclables in the case of materials recovery facilities); hence a source of revenue to meet capital and operating costs. Flow controls are highly controversial and often fought, legally, on the basis that they limit free commerce (especially interstate).

In addition to local flow controls, competition for MSW may be found in tipping fees charged for local waste versus waste brought in from outside the area of operation. For example, 1992 tipping fees for local waste were reported to be $55 and $63.50 in Broward County, Florida and Montgomery County, Pennsylvania, respectively. Tipping fees for outside waste, in these counties, was reported at $42 and $41 per short ton. Both localities were also reported to use flow controls to guarantee local waste (EPA-530-R-95-008 1995).

Finally, tipping fees are affected by the manner in which the MSW is disposed. For example, the tipping fees for MSW that was disposed in landfills averaged about $34.29/ton nationwide (in 2004) while the tipping fees for MSW that was incinerated averaged about $61.64 nationwide (Repa 2005). This difference can be attributed to the increased cost of incineration as a means for disposal. By adding a processing cost to the tipping fee for MSW disposal, it may be possible to reduce the scale required for converting MSW to liquid fuels to a smaller size that better fits the MSW disposal site capacity. A more detailed economic analysis would be required to determine how much processing cost would need to be added to a tipping fee for various plant sizes, along with an assessment of the willingness of the public to accept these additional charges.

4.0. AVAILABILITY AND DISTRIBUTION OF MSW

MSW has an immediate advantage over biomass as a feedstock in that it is already collected at sites around the country and disposed in landfills. However, there is considerable variability in the annual capacity of these sites and, as was previously discussed, there is probably a minimum capacity below which the economics are not favorable for converting MSW into liquid fuels.

MSW generation rates are often given per capita to understand distribution, and are based on data compiled by the EPA, further supported by the U.S Census Bureau. Thus, a map showing population density would be indicative of the distribution of the nation's MSW stream. Figure 4.1 is a population density map generated using U.S. Geological Survey and U.S. Census Bureau data (National Atlas of the United States). Note that the Mid-West and Rocky Mountain regions show the least population density. Interestingly, several states in this region allow back-yard and mass burning of MSW. In any case, supplying sufficient feedstock to a thermochemical conversion plant appears to be most likely in regions on the East and West coasts.

Actual availability of MSW, within a given region, is dependent upon several, dynamic factors. The implementation of the RCRA Subtitle D regulations in 1993 spurred a movement from small, distributed landfills to larger landfills equipped to contain contamination and vent combustible gases. This trend toward larger landfills is shown in Figure 4.2 (Franklin Associates 2008). At the same time landfill site management shifted towards large companies with landfill site management expertise to implement the RCRA regulations. For example, in 1992, there were 5386 landfill sites reported by the EPA in the United States, of which 348 sites were owned or operated by the 13 largest waste management companies that operated landfill sites (approximately 6% of all active sites) (EPA-530-R-95-008 1995). By 2008, the EPA reported only 1754 active landfill sites. An Internet search of the major waste management companies indentified 574 landfill sites owned or operated by the 13 largest waste management companies that operated landfills (approximately 33% of all active sites). The trend towards fewer larger landfill sites accompanied by the trend towards more sites operated by waste management companies, who could be important collaborators, should improve the potential for finding opportunities for siting MSW liquid fuels synthesis plants.

A MSW mass range of 3,200-3,400 short tons per day (described above in Section 3.2) was determined to be equivalent to the economically-viable 2,000 tonne per day wood gasifier, based on relative moisture contents and heating values (Figures 3.1 and 3.2). Landfill disposal weights (or volumes[2]) were collected from 44 states[3,4]. This data was inspected to evaluate the number of landfills, nation-wide, meeting a 3,300 short tons per day criteria. The results are shown in Figure 4.33

These preliminary results indicate 47 potential landfill sites receiving 3,300, or more, short tons per day.

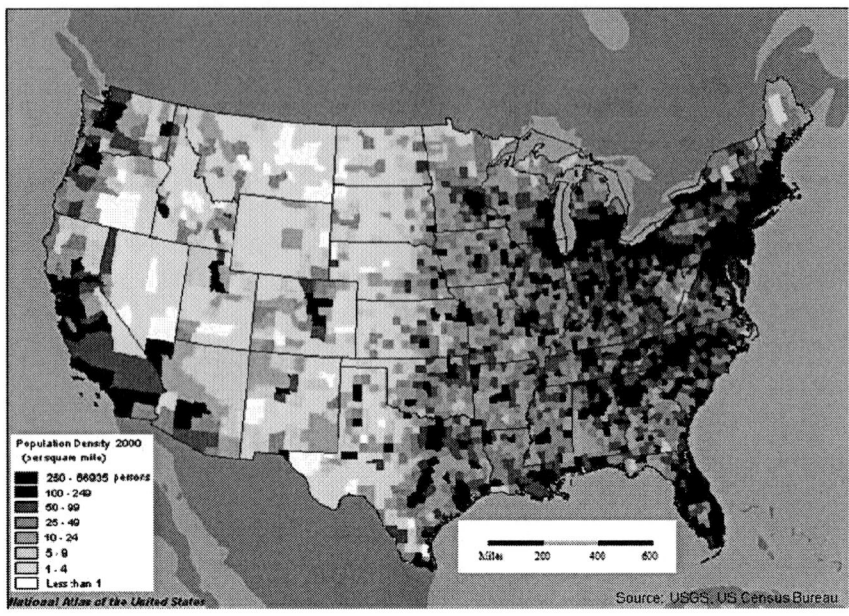

Figure 4.1. United States Population Density (Interior 2008).

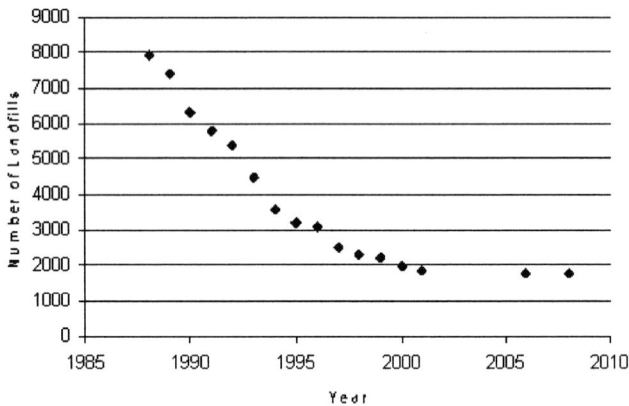

Figure 4.2. Number of Landfills, Nation-Wide, Since 1988.

Interestingly, these 47 landfills take in an average of 5,700 short tons per day, each. For perspective, assume that MSW gasification can support a liquid fuels (e.g. ethanol) synthesis operation at a rate of approximately 48.6 gallons per short ton (roughly 2/3 the potential 72.6 gallons of ethanol per ton from wood gasification, based on relative heating values). If 47 plants processed

5,700 short tons per day, and generated 48.6 gallons of ethanol per short ton, the expected output of the combined plants would be 310,000 bbl per day, or 113 MM bbl per year. The EIA reported a total liquid fuels consumption, in the U.S., of approximately 40 quadrillion Btu in 2006 and (as shown in Figure 1.1) a transportation sector demand of 28 quadrillion Btu in 2008. Assuming a HHV of 84,000 Btu per gallon of ethanol, 47 plants could supply 0.4 quadrillion Btu in 2006 (1% of total liquid fuels consumption) or 1.4% of transportation sector energy demand.

The daily throughput of landfills, as collected from the 44 states in which site-specific data was available, is shown in Figure 4.4. Although 90% of the nation's landfills process 1750 short tons per day or less, almost 30% of MSW goes to landfills processing 3,300 short tons per day or more. An additional 30% of the nation's MSW enters landfills handling between 1250 and 3300 short tons per day. Thus, an economically viable gasification technology, operating at these smaller scales, could be important.

It should be noted that the data collected from local government departments on individual landfills (or counties) differed greatly from the numbers reported by larger entities, such as the EPA. The total MSW entering MSW landfills, as reported by local governments, was 341 MM short tons. This is more than 35% greater than the 251 MM short tons produced in the US, based on EPA estimates. Furthermore, the EPA estimates that a significant portion of the produced MSW is recycled or composed, thus not entering landfill sites. The higher estimate based on local governments and local landfills may be due to many factors including non-MSW (C & D debris, biosolids, and agricultural wastes) material entering landfills, the use of estimation at landfill sites without scales, and the uncertainty of product lifetimes assumed in the material balance method used by EPA. Other factors adding to this discrepancy involve uncertainties regarding diversions that occur after material has entered a given landfill. For example, a landfill site may report MSW entering the landfill gate that is later sorted for recycle, or used in an incinerator or composting operation instead of being disposed in a landfill. The difference between estimated MSW generation and state reporting has also been documented in other 2004 and 2006 studies (Simmons 2006, 2006a). This further justifies the necessity for careful data evaluation.

It should also be noted that large MSW landfill sites are already generally more technologically advanced. This advancement was likely set in motion by RCRA Subtitle D, which limits landfill gas concentrations of methane in occupied structures onsite and soil atmosphere at the property line of the landfill site to 1.25% and 5%, respectively. Preliminary investigation into the potential landfill sites revealed that 58% already have landfill gas (LFG) recovery operations in place. Other sites incinerate MSW to produce electric power. This means that a MSW gasification strategy may come to be in direct competition with these existing operations. On the other hand, LFG recovery and gasification systems could be complementary, with the methane used to raise steam for the gasification plant, ensuring that more of the MSW goes to liquid fuels. Such a cooperative venture would be temporary, unless carefully strategized to maintain sufficient LFG generation.

5.0. CURRENT STATE OF GASIFICATION TECHNOLOGY

There are more than 140 gasification plants currently in operation worldwide comprising more than 420 gasifiers. Most of these gasifiers use coal feedstocks. Worldwide gasification capacity is expected to increase 80% by 2015 with most of that increase occurring in China (Gasification Technologies Council 2008).

Biomass gasification and incineration are proven commercial technologies worldwide for both conventional biomass and MSW. However, these applications are generally limited to firing boilers for process heat and electricity generation and, in a few cases for gasification supplying a gaseous fuel to a diesel generator system. The process scale for these applications is case specific ranging anywhere from a few to several hundred dry tons/day with applications for power generation generally in the larger sizes. There have been some demonstration-scale efforts with conventional biomass feedstocks to provide gas for a gas turbine, but inadequate reduction of tar levels in the product gas has proven to be a technical barrier at this time (Cobb 2007). Biomass gasification for liquid fuels synthesis using either conventional biomass feedstocks or MSW has not been developed at this time. This is mainly due to economic issues involving maximum practical scale of biomass conversion plants with respect to resource availability and the minimum scale required to produce liquid fuels using synthesis gas for various liquid products. Closely tied to these economic issues are technical issues regarding efficient and cost effective gas cleanup technologies to remove tars and trace gas contaminants such as H_2S as well as efficient catalystic processes to convert synthesis gas into liquid products.

Figure 4.3. U.S. Landfills Receiving ≥ 3300 Short Tons per Day.

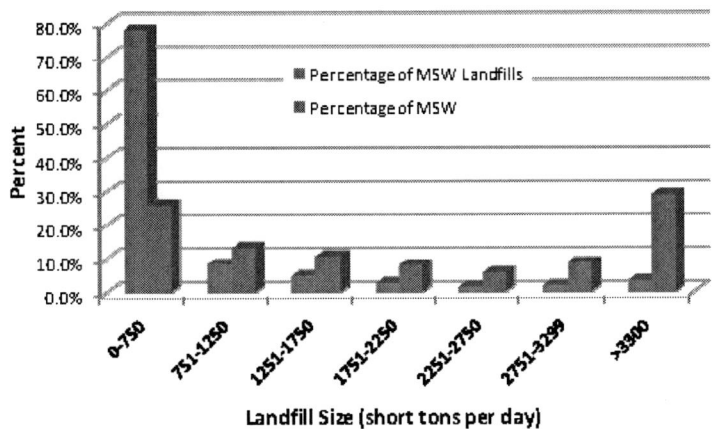

Figure 4.4. Throughput Distribution Processed by the Nation's Active Municipal Solid Waste Landfills.

There is considerable variety in the specific designs of gasifiers. Many of these variations are derived from specific application needs regarding process scale, gas quality, feedstock introduction, and ash management. However, from a flow dynamics perspective, there are four general gasifier types—fixed bed, fluidized bed, entrained flow, and plasma reactors. Each general gasifier type has unique benefits and drawbacks depending not only on the desired feedstock but the end product as well.

5.1. Fixed Bed Gasifiers

The fixed bed gasifiers are relatively simple in design with biomass added to the top of the gasifier vessel and ash and unreacted char removed from the bottom. They are called fixed bed because the biomass does not move freely in the reactor. Instead the biomass moves slowly downward through a fixed zone supported by a grate at the bottom of the gasifier. Fixed bed gasifiers are relatively small in size and are directly heated using air, although purified oxygen could also be used.

There are two main fixed bed gasifier designs. An up-draft gasifier is a counter flow design in which biomass moves downward through the gasifier while air is introduced into the bottom of the gasifer and the reacted gases exit the top. In this configuration, biomass entering the top is slowly heated as it moves down through the bed giving off moisture, pyrolysis gases (hydrogen, carbon monoxide, carbon dioxide, and light hydrocarbon gases), and vaporized pyrolysis liquids called tars that are carried upward with the gas flow leaving with the product gas at concentrations typically ranging from 1.2×10^{-3} to 5.9×10^{-3} lbs/scf (NREL/TP-570-25357). Gases exiting the updraft gasifer are at a relatively modest temperature of approximately 930°F (Ciferno and Marano 2002). Char produced from slow pyrolysis of the biomass continues downward to a grate where air is introduced combusting the char. Heat from this combustion process drives the endothermic drying and pyrolysis reactions. Ash and unreacted char drop through the grate and are removed from the gasifier. A simplified illustration of an up-draft gasifier is given in Figure 5.1.

In a down-draft gasifier design, biomass also is added at the top of the reactor and works its way downward, slowly drying and pyrolyzing. The moisture, pyrolysis gases, and tars produced as the biomass slowly pyrolyzes pass downward to the bottom of the gasifer vessel passing through a zone where air is introduced to the gasifier. The air combusts the pyrolysis gases, tars, and a portion of the char producing a high temperature zone (approximately 1800°F), where steam and CO_2 in the gas stream can react with the remaining char producing additional hydrogen and carbon monoxide (Cirferno and Marano 2002). The gas stream leaves the gasifier just above a grate at the bottom of the vessel at approximately 1475°F, while ash and unreacted char fall through the grate and are removed. Tars in the product gas typically range from 5.9 x 10^{-6} to 7.1 x $1\,0^{-5}$ lbs/scf (Milne et al. 1998). A simplified illustration of a down-draft gasifier is given in Figure 5.2.

There are hybrid fixed bed gasifier designs that possess elements of updraft and downdraft gasifiers. Baffles and alternative points of air introduction create combustion zones that may not be fully mixed with all gases and tars evolved in cooler parts of the gasifier as would occur in a downdraft gasifier. However, those gases and and tars do experience much higher temperatures than are experienced in an updraft gasifer, providing opportunities for tar cracking.

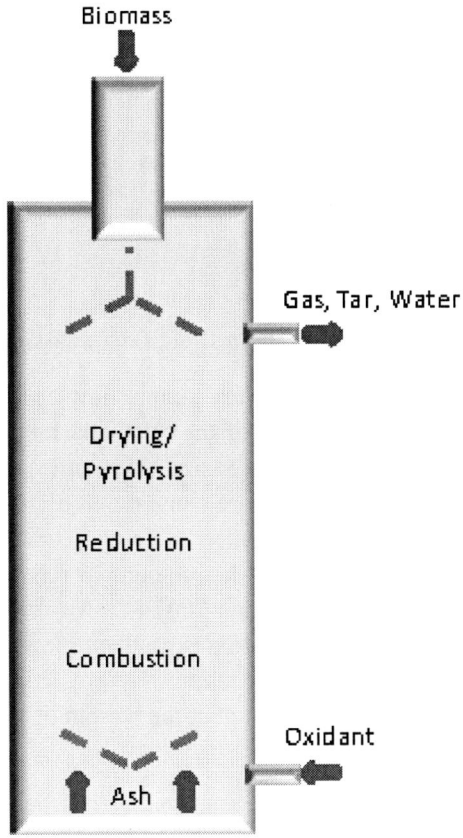

Figure 5.1. Simplified Schematic of an Up-Draft, Fixed-Bed Gasifier.

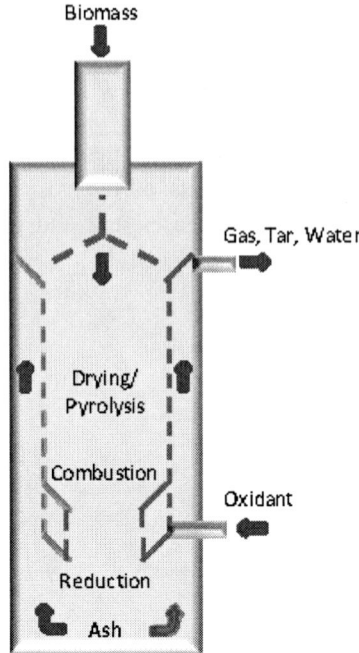

Figure 5.2. Simplified Schematic of a Down-Draft, Fixed-Bed Gasifier.

Both fixed bed gasifiers designs are simple, low-cost processes but they are also limited in capacity because of the relatively long biomass residence times and practical limits on vessel size. Up-draft gasifiers can handle feedstocks with relatively high moisture content and high ash content making them suitable for MSW with minimal processing. Their main disadvantage besides size is the high tar content in the product gas reaching as high as 10% to 20% by weight (Ciferno and Marano 2002). Down-draft gasifiers are very efficient in converting pyrolysis tars to gas producing a gas that is essentially tar-free, at least in smaller-scale applications. Both gasifiers can handle fairly coarse material although the constriction in the bottom of a down-draft gasifier is more limited in this aspect.

5.2. Fluidized Bed Gasifiers

Fluidized bed gasifiers are characterized by the use of a heated solid medium, such as sand made to bubble (or "boil") by introducing the gas stream below the sand bed at sufficient velocity to barely lift (fluidize) the sand so that the sand particles are free to move throughout the sand bed. Biomass is introduced into the fluidized bed either from the top or from the side and mixed with the hot sand causing rapid pyrolysis to produce mostly gas and a small quantity of tar and char. A direct fired fluidized bed is heated by introducing air or purified oxygen into the bottom of the bed as all or part of the fluidizing gas (steam is also added in some applications). The oxygen in the fluidizing gas combusts a portion of the pyrolysis gases, char and tar then passes out the top of the gasifier vessel. Char is ground up by the motion of the sand to small char and ash particles that are carried out of the gasifier where they must be removed by cyclones. An indirectly heated fluidized bed gasifier withdraws a portion of the fluidized bed contents including the char in the bed and transfers it to a second fluidized bed where air is introduced into the bottom of the second vessel to combust the char thereby heating the sand to a high temperature. The hot sand is then transferred back to the gasifier where the hot sand is used to heat the biomass. Only steam is introduced to the bottom of the gasifer in this configuration to fluidize the bed. This type of configuration is often called a circulating fluidized bed because the solids are withdrawn and reintroduced to the gasifer. Fluidized bed gasifiers typically operate at moderately high temperatures (1300°F–1750°F), utilize a solid fuel stream, and have residence times ranging from one second to a minute. Simplified illustrations of a fluidized bed gasifier and a circulating fluidized bed gasifier are given in Figures 5.3 and 5.4, respectively.

Fluidized bed gasifiers can be scaled to relatively high capacities because of the much shorter biomass solids residence times in the gasifier. However, there are more limitations on the physical properties of the feedstock because the solids must be fluidizable or they must be removed from the bottom of the vessel. In the case of MSW, this would most likely require separation of large heavy objects (metals and other heavy inorganic solids) and size reduction and screening to achieve particles with similar aerodynamic properties with respect to fluidization. RDF would be particularly suitable as a feedstock. Product gas from fluidized bed gasifiers also contains a small quantity of tars typically ranging from $5.9 \times 10^{-5} - 8.9 \times 10^{-4}$ lbs/scf (Milne et al. 1998). These tars must be reduced to much lower levels for synthesis gas applications, either by separation or by conversion into gas.

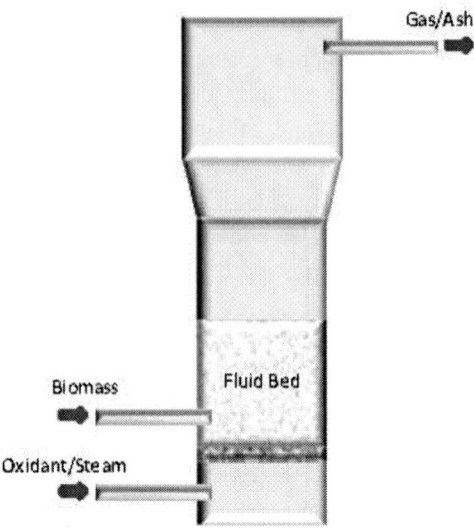

Figure 5.3. Simplified Schematic of a Fluidized Bed Gasifier.

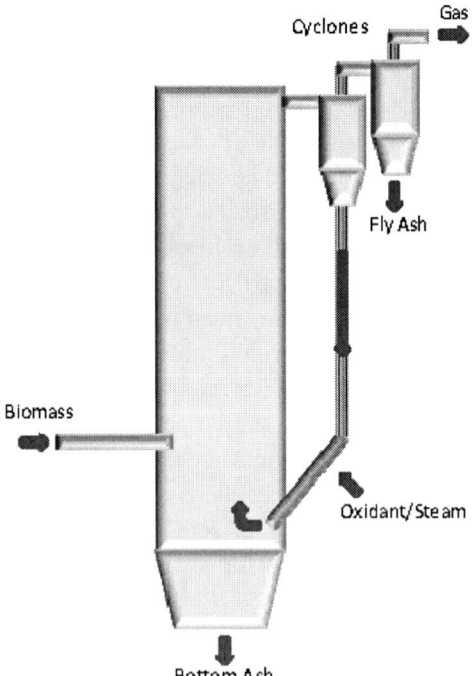

Figure 5.4. Simplified Schematic of a Circulating, Fluidized Bed Gasifier.

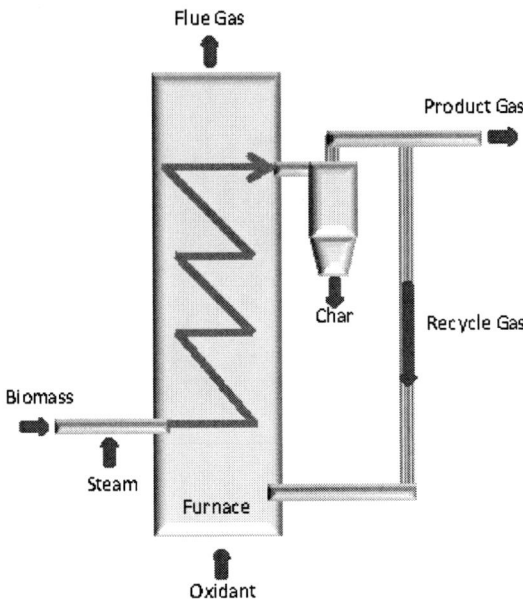

Figure 5.5. Simplified Schematic of an Entrained Flow Gasifier.

5.3. Entrained Flow Gasifiers

Entrained flow gasifiers (also sometimes called circulating bed gasifiers) are similar to fluidized bed gasifiers except that the gases introduced into the bottom of the gasifier vessel are at a much higher velocity causing the fluidized medium and biomass to become entrained and carried out the top of the gasifier. Solids residence times in the gasifier range for 1-10 seconds and the temperatures are higher (1650°F–2550°F). Solids are removed from the product gas and either returned to the gasifier or sent to a separate combustor where the char is combusted. An entrained bed gasifier may be directly fired with air or oxygen or indirectly heated (using the separate combustor to heat the sand). The advantages and disadvantages of entrained flow gasifiers are similar to those for fluidized bed gasifiers except the throughputs can be greater, but the solids properties are usually more stringent. The high velocities of the entrained solids may also accelerate equipment erosion, compared to a fluidized bed gasifier. Figure 5.5 is a simplified illustration of an entrained flow gasifier.

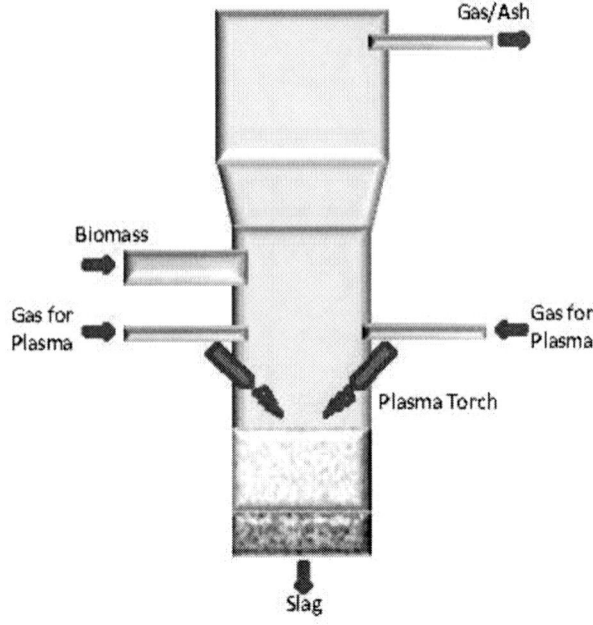

Figure 5.6. Simplified Schematic of a Plasma Gasifier.

5.4. Plasma Gasifiers

Plasma gasifiers are characterized by the use of plasma, a low-density ionized vapor, to heat the feed stream. Air, nitrogen, argon, carbon dioxide, and steam may be used as the plasma gas. Feedstock materials are treated using extremely high temperatures (1700°F–35,530°F) that convert solid or liquid fuel streams into synthesis gas and vitrified slag. In general, thermal plasmas may be generated by either an electric arc or by a radio frequency induction discharge. Arc plasmas are typically used for waste treatment because they are relatively insensitive to changes in process conditions. Further reaction may be enabled by a molten metal or glass layer. This configuration is referred to as a plasma melter. Plasma gasifier technologies are considered to be well established, particularly for vitrification of incinerator ash. Electric power requirements for treating waste[6] have been reported in the range of 0.34–4.4 MM Btu per ton of waste (Heberlein and Murphy 2008). Figure 5.6 shows a simple illustration of a plasma gasifier.

6.0. CURRENT STATE OF MSW GASFIERS

Gasifiers currently using MSW feedstocks are shown in Table 6.1. Information compiled in this table is based on information available in the literature (New York City 2008, Los Angeles County 2005, Westinghouse Plasma Corporation 2007, 2007a) and the gasifier sponsor's response to requests for information. The MSW gasifiers are grouped according to their status of development. Many of these designs are focused on producing heat and electric power and, consequently, are designed with close coupled combustion of the gases and using air as the oxidant in the gasification and combustion sections of the process. These designs would require modifications to the design or operation in order to produce a product gas suitable for liquid fuels synthesis. In a couple of cases, fuel gas or synthesis gas, for methanol synthesis is indicated. In these cases, minimal design changes would be expected. Coincidentally, those indicating methanol synthesis also operate in a manner where MSW is co-fed with coal.

6.1. Commercial Gasifier Descriptions

Four commercial gasifiers were identified that process MSW. These are described below.

6.1.1. Westinghouse Plasma
Westinghouse plasma torches were originally built for metallurgical applications. Westinghouse also builds a plasma gasification vitrification reactor system, which combines their torch with a moving bed. There are two commercial waste processing facilities and one commercial ash vitrification plant in Japan, all established between 1995 and 2002. One waste processing facility is a 242 short ton per day mixed auto shredder and MSW operation for power generation. The other is a 30 short ton per day mixed MSW and sewage sludge for heat production, which is then used in the waste water treatment facility (Westinghouse Plasma Corporation 2007). Westinghouse Plasma Corporation claims that capacities between 500 and 750 tons/day can be handled in a single reactor vessel (Westinghouse Plasma Corporation 2007a). The torch can be oxygen-blown and the company claims that it can make synthesis gas as a product (Westinghouse Plasma Corporation 2007a).

Table 6.1. MSW Gasifiers Currently in Operation

Sponser	Gasifier	Type	Primary Feedstocks	Capacity (tpd)	Pretreatment	Products	Gas Treatment
Commercial Scale							
Rigel Resource Recovery	Westinghouse Plasma	Plasma	MSW, specialty wastes	72-300	Potential sizing	Power	Scrubbing
Ebara	Twin-Rec	Circulating fluidized bed + slagging ash melter (both air blown)	Auto shredder waste, MSW, sewage sludge	462 (modular, 154 ea)	Sizing to 300 mm	Electricity, building materials	
Sekundarrohstoff-Verwertungszentrum	Schwarze Pumpe	O$_2$ blown entrained flow (GSP) + slagging (BGL) + pressurized bed-rotating grate (FDV-Lurgi)	Lignite, tar oils and waste	2030	Sizing, briquetting, pelletizing	Fuel gas/methanol	Rectisol
Global Energy solutions, LC	Thermal Converter	3-phase, 2-chamber downdraft: pre-heated air (400°C), primary conversion (1300°C), secondary conversion in molten bed (1675°C)	MSW	24-600 (modular)	None	Heat/power	
Demonstration Scale							
Krupp Uhde GmbH	High Temperature	Fluidized bed (O$_2$, steam blown)	MSW co-fired with coal	660	Pelletizing	Methanol ls	
Interstate Waste Technologies	Winkler Thermoselect	Closed-loop, high T, O$_2$/NG blown	MSW	110-792 (modular, up to 396 tpd ea)	Compaction, degassing	Heat/power, liq fuels	Quench, S, heavy metal

6.1.2. Twin-Rec

The Twin-Rec is a circulating fluidized bed system coupled with a slagging ash melter. MSW is shredded and screened to 12 inches or less and fed into the air-blown, fluidized bed gasifier, which contains a fluidizing medium (sand) and operates at 930°F–1200°F (Hotchkiss et al. 2002). Heavy inert material, such as glass, ceramics, and metals, as well as some of the ash, is extracted from the bed along with some of the sand. The sand is mechanically separated from the other materials and returned to the fluidized bed. Fine char, fly ash, and fuel gases, generated in the fluidized bed, carried over into the cyclonic melting chamber where the gases and char are combusted in air in a at 2,400°F–2,600°F. The hot gases leaving the cyclonic melting chamber are fed into a boiler to produce steam for electricity generation and process heat. Molten slag leaving the bottom of the cyclonic melter chamber is quenched to form granules. Both the gasifier and melter operate at atmospheric pressure, without auxiliary fuel (except during start-up) or oxygen. The system requires a steam generator. The waste materials are converted to electricity and/or heat. Ebara (Ebara 2007) claims 14 commercial Twin-Rec facilities treating waste plastics, shredder residue, sludges, industrial waste and MSW.

6.1.3. Schwarze Pumpe

Schwarze Pumpe is a location in Germany that houses one of the largest German plants handling solid and liquid wastes including plastics, MSW, sewage sludge, shredder residues, oils, tars, and solvents. Sekundärrohstoff VerwertungsZentrum (SVZ) is a large scale operator of three types of gasifiers at the location—entrained flow (GSP), slagging (BGL), and rotating grate (FDV, Lurgi). The site opened in 1955 as a briquetting factory and power station. Schwarze Pumpe began generated town gas in the late 1960s. Their experience with MSW started in the mid-1990s with an emphasis on methanol production (Kamka and Jochmann 2005).

6.1.4. Thermal Converter

Global Energy Solution's Thermal Converter resembles a slagging, down-draft gasifier. MSW is shredded and screened to a lumped size no greater than a three inch cube. It has three distinct sections. Preheated air is introduced into the top of the fixed bed to heat and dry the MSW in the upper portion of the vessel. The middle section of the vessel is a rotated section where MSW is pryolyzed at temperatures ranging from 2,600°F -2,730°F. In the lower section of the vessel, the inorganic fraction of the MSW is converted into a molten bed of slag that drips from the bottom of the vessel through an after-burner section and into a quench tank where the slag is cooled and solidified. Gases passing through the molten slag enter the after-burner where additional air is introduced to convert any remaining combustible gases. This zone reaches temperatures of 2,900°F. Heat is recovered from the hot exhaust gases in a heat exchanger where inlet air is preheated and a recuperator where, presumably, steam can be produced for electric power generation and process heat. Preheated air for the gasifier can also be produced in the recuperator (Kamka and Jochmann 2005). This gasifier, as currently configured, is closely coupled with a combustion zone (afterburner) that precludes the production of a syngas, even if purified oxygen were to be used instead of air. The largest unit appears to have a capacity of about 75 tons/day.

6.2. Demonstration Gasifier Descriptions

Two demonstration-scale gasifiers were identified. These are described below.

6.2.1. High Temperature Winkler Gasifier

The High Temperature Winkler (HTW) plant in Berrenrath, Germany is commercial (for lignite) and has been in operation since 1986. Dried lignite, oxygen, and steam are fed to a fluidized bed reactor that operates at 145 psi and1750°F. Product gas from the gasifier is cooled and cleaned in a ceramic candle filter and a water scrubber. Solid residues are combusted in an adjacent power plant. Dust collected from the filter is used in an adjacent wastewater treatment plant. The synthesis gas undergoes carbon monoxide conversion to obtain a hydrogen to carbon monoxide ratio suitable for methanol synthesis. Hydrogen sulfide and carbon dioxide are removed. The resulting syngas is sent to DEA Mineralol AG where it is synthesized into methanol. A series of tests were performed, using the HTW system, to successfully co-gasify RDF pellets (at rates up to 50% of the feed input) in 1998. The demonstration was completed that year (Hotchkiss et al. 2002).

6.2.2. Thermoselect Gasifier

The Thermoselect Gasifier handles raw MSW. The feedstock is preshredded to 20 inches or less, compressed into plug that is ram fed into a long degassing chamber that moves the MSW horizontally to the gasification chamber that is vertically oriented. Heat radiated from the gasification chamber slowly pyrolyzes the compacted MSW as it slowly moves through the degassing chamber (approximately 1 to 1.5 hours solids residence time). Gases and vapors produced in the degassing chamber pass upwards through the gasification chamber where they are exposed to temperatures of approximately 2190°F for at least two seconds. Char and inorganic solids enter the gasification chamber and drop to its bottom, where the char and supplemental natural gas are combusted with purified oxygen producing temperatures as high as 3600°F causing the inorganic materials to form a molten slag. The molten slag, consisting of separate metal and mineral layers, is quenched in a water quench basin forming granules. The granules are sorted according to their properties to recover the metals. The product gas is quenched and cleaned in a multi-stage system in preparation for power generation or chemical synthesis (City of Los Angeles Department of Public Works 2005). Thermoselect's pilot plant was operated in Italy. There are 7 facilities in Japan, three of which handle MSW. The remaining 4 facilities gasify industrial waste. Two facilities export fuel gas. All facilities handling MSW use the fuel gas for gas engines. The MSW facilities use 2-3 lines to treat 120-300 tons per day (Thermoselect 2008).

CONCLUSIONS

MSW is a potential gasifier feedstock that presents an opportunity to produce alternative liquid fuels because of its availablility in significant amounts at current landfills and because it is a predominantly biomass derived material that, like conventional biomass feedstocks such as wood, is a renewable resource. The viability of MSW as a gasifier feedstock for liquid fuels synthesis depends on several factors. Foremost is the availability of MSW in sufficient quantities to meet the minimum process scale required for economic feasibility. Based on this review, that process scale may be as large as 3,300 short tons per day of as received MSW, based on the anticipated scale required for conventional biomass feedstocks. However, the required scales for both types of feedstock depend on the cost of the feedstock, which has a significant effect on process economics. Conventional biomass feedstocks are market based and are a cost to the process, averaging nearly $45 per dry short ton, whereas MSW is charged to the supplier as a tipping fee to dispose of the material. Historically, landfills receiving MSW and processing it to produce heat and energy using incinerators and gasifiers charge about $30 per short ton (as received basis) additional fee beyond that typically charged to just landfill the material. The effect of the extra charges on process economics and, in turn, minimum process scale for economic feasibility needs to be further examined.

A review of available information on the number and size of various landfill sites around the country identified 47 sites that processed 3,300 short tons per day or more (as received basis) of MSW. Together these sites could potentially produce enough liquid fuel to meet approximately 1.4% of current transportation fuel demand (about 113 MM bbl/year of liquid fuel). A greater contribution could be attained if smaller scale facilities are found to be feasible due to latitude in the tipping fee charged to MSW producers.

Another important issue deals with the quality of MSW as a feedstock. MSW is a heterogeneous feedstock containing materials with widely varying sizes, shapes, and composition, which can lead to variable gasification behavior if used in an as received condition. It is expected that some minimal size reduction and sorting will need to be performed to make MSW suitable as a feedstock for MSW gasifiers. RDF is a processed form of MSW where significant size reduction, screening, sorting and, in some cases, pelletization is performed to improve the handling characteristics and composition of the material to be fed to a gasifier. There is a trade-off between the increased costs of producing RDF from MSW and potential cost reductions in gasifier design and operation.

The chemical make-up of MSW includes significant quantities of chemical constituents that can create problems in downstream processes. While the concentrations of these contaminants are greater than those found in conventional biomass feedstocks, they are roughly comparable to those found in coal. The commercial operation of 2 gasifiers co-feeding coal and MSW for methanol synthesis suggests that gas cleanup to remove key contaminants can be accomplished with existing technology.

This study identified four commercially available MSW gasifiers and two demonstration gasifiers. Three of these gasifiers produce only electricity and process heat so they are closely coupled to combustors. Consequently, they would require some design changes to be adapted to producing a synthesis gas. The commercial and demonstration gasifiers are currently available in sizes that range from 24 to 660 short tons per day MSW processing capability and will likely require multiple gasifiers to meet the minimum processing scale requirement for a liquid fuels synthesis plant. Further investigation is needed to determine the trade-offs between using many relatively small scale gasifiers that may be built as packaged systems or a few larger field erected gasifiers to minimize gasifier capital and operating costs. In addition, there are a large number of gasifier designs with a range of capacities that are at the pilot scale level of development. These were not examined closely and may ultimately be suitable for syngas applications.

Overall, this study concludes that MSW should be considered as a potentially viable gasifier feedstock for liquid fuels synthesis. A review of feedstock availability, composition, and handling characteristics, along with commercially available MSW specific gasifiers, did not identify any obvious insurmountable technical or economic barriers to commercialization. However, further research into the economic issues surrounding tipping fees and process scale is needed to verify economic viability and the appropriate plant scale for economic viability.

REFERENCES

340-04-005. (2004). *Statewide Waste Characterization Study*. California Integrated Waste Management Board. Sacramento, California.

40 CFR Part 257. *Criteria for Classification of Solid Waste Disposal Facilities and Practices*

40 CFR Part 258. *Criteria for Municipal solid Waste Landfills*

California Biomass Collaborative. (2006). *An Assessment of Biomass Resources in California, 2006*. Department of Biological & Agricultural Engineering, University of California. Davis, California.

California Biomass Collaborative. (2007). *Municipal Wastes-Background Discussion Paper*. Williams RB. Department of Biological & Agricultural Engineering, University of California. Davis, California.

Caputo, A. C. & Pelagagge, P. M. (2002). "RDF Production Plants: I Design and Costs." *Applied Thermal Engineering. Vol. 22*, Issue 4. Pages 423-437. Elsevier Inc., Burlington, Massachusetts. Accessed June 27, 2008 at http://dx.doi.org/ 10.1016/S1359-4311(01)00100-4.

Ciferno, J. P. & Marano, J. J. (2002). Benchmarking Biomass Gasification Technologies for Fuels, Chemicals, and Hydrogen Production. E²S for *National Energy Technology Laboratory*.

City of Los Angeles Department of Public Works. (2005). Appendix E, Supplier Evaluations, in *Evaluation of Alternative Solid Waste Processing Technologies*. Prepared by URS Corporation for the City of *Los Angeles Department of Public Works*. Los Angeles, California. Accessed

September 29,2008 at http://www.lacity.org/san/solid_ resources.

Cobb, J. (2007). *Production of Synthesis Gas by Biomass Gasification –A Tutorial.* 2007 *AIChE Spring National Meeting.* April 22-26, 2007.

DOE/EIA-0383. (2008). *Annual Energy Outlook 2008.* Energy Information Administration, US Department of Energy, Office of Integrated Analysis and Forecasting. Washington, *District of Columbia.*

Ebara. (2007). *TwinRec – Fluidized Bed Gasification and Ash Melting.* Ebara. Zurich, Switzerland. Accessed September 18, 2008 at *http://www.ebara. ch/_en_/twinrec.php?n=1.*

Ecology. (2006). *Solid Waste In Washington State 14th Annual Report.* Solid Waste and Financial Assistance Program, Washington State *Department of Ecology.* Olympia, Washington.

Ecology, (2008). Solid Waste and Recycling Data. Washington State Department of Ecology. Olympia, Washington. Accessed May 13, 2008 at http://www.ecy.wa.gov/programs/swfa/solidwastedata/.

Ecology. (2008a). "Tipping Fees at MSW Landfills." *Solid Waste and Recycling Data.* Washington State Department of Ecology. Solid Waste and Financial Assistance Program. Olympia, Washington. Accessed June 27, 2008 at http://www.ecy.wa.gov/programs/swfa/solidwastedata/ disposal/TippingFees.xls.

EPA. (2007). *Municipal Solid Waste in the United States: 2006 Facts and Figures.* Office of Solid Waste (5306P), US Environmental Protection Agency. Washington, DC.

EPA. (2008). *MSW Characterization Methodology.* Office of Solid Waste (5306P), United States Environmental Protection Agency. Washington, District of Columbia. Accessed June 13, 2008 at *http://www.epa.gov/ epaoswer/non-hw/muncpl/pubs/06numbers.pdf.*

EPA-530-F-07-030. (2007). *Municipal Solid Waste in the United States: Facts and Figures for 2006.* Office of Solid Waste (5306P), United States Environmental Protection Agency. Washington, *District of Columbia.*

EPA-530-R-95-023. (1995). *Decision-maker's Guide to Solid Waste Management 2nd Edition.* Office of Solid Waste, RCRA Information Center (5305W), United States Environmental Protection Agency. Washington, District of Columbia.

EPA-530-R-95-008. (1995). *Flow Controls and Municipal Solid Waste.* Office of Solid Waste, Municipal and Industrial Solid Waste Division, United States Environmental Protection Agency. Washington, District of Columbia.

EPA-530-R-99-009. (1999). *Biosolids Generation, Use, and Disposal in The*

United States, Solid Waste and Emergency Response (5306W), United States Environmental Protection Agency. Washington, District of Columbia.

EPA-530-R-99-013. (1999). *Cutting the Waste Stream in Half: Community Record-Setters Show How.* Solid Waste and Emergency Response (5306W), United States Environmental Protection Agency. Washington, *District of Columbia.*

EPA-530-R-99-016. (1999). *Organic Materials Management Strategies.* Solid Waste and Emergency Response (5306W), United States Environmental Protection Agency. *Washington, District of Columbia.*

EPA-832-R-06-005. (2006). *Emerging Technologies for Biosolids Management.* Office of Wastewater Management, United States Environmental Protection Agency. Washington, District of Columbia.

Florida Department of Environmental Protection. (1999). "Chapter 4: Landfill Disposal," *1999 Solid Waste Management Annual Report.* Bureau of Solid & Hazardous Waste, Division of Waste Management, Florida Department of Environmental Protection. Tallahassee, Florida.

Franklin Associates. (2008). *Integrated Solid Waste Management and Planning.* Accessed 13 June 2008 at http://www.fal.com/solidwaste.htm.

Gasification Technologies Council. (2008). "State of the Gasification Industry," *What is Gasification?* Gasification Technologies Council. Arlington, Virginia. Accessed August 12, 2008 at http://www.gasification

Hotchkiss, R., Livingston, W. & Hall, M. (2002). *Waste/Biomass Co-Gasification with Coal.* Report No. COAL R216. DTI/Pub URN 02/867. Cleaner Coal Technology Transfer Programme, Department of Trade and Industry, United Kingdom.

Interior. (2008). "Population Density 2000 Map," National Atlas of the United States. United States Department of the Interior. Washington, *District of Columbia.* Accessed 28 July 2008 at: *http://nationalatlas.gov/natlas/ Natlasstart.asp.*

Kamka, F. & Jochmann, A (2005). *Development Status of BGL Gasification.* International Freiberg Conference on IGCC & XtL Technologies.

Klass, D. L. (1998). *Biomass for Renewable Energy, Fuels, and Chemicals.* Academic Press, A division of Elsevier Inc. Burlington, Massachusetts.

Los Angeles County. (2005). *Conversion Technology Evaluation Report.* URS Corporation for the Alternative Technology Advisory Subcommittee, Los Angeles County Integrated Waste Management Task Force. Los Angeles, California.

New York City. (2008). "Phase I: Evaluation of New and Emerging Solid

Waste Management Technologies." *Evaluation of New and Emerging Waste Management and Recycling Technologies and Approaches.* Economic Development Corporation and Department of Sanitation. New York, New York.

NREL/TP-42 1-7501. (1994). *Review and Analysis of the 1980-1989 Biomass Thermochemical Conversion Program.* Stevens DJ. National Renewable Energy Laboratory. Golden, Colorado.

NREL/TP-43 1-4988A. (1992). *Data Summary of Municipal Solid Waste Management Alternatives; Volume I.* SRI International for National Renewable Energy Laboratory. Golden, Colorado. Accessed June 27, 2008 at http://lists.p2pays.org/ref/11/10516/.

NREL/TP-43 1-4988A. 1. 1992. "Refuse-Derived Fuel." *Data Summary of Municipal Solid Waste Management Alternatives; Volume I.* SRI International for National Renewable Energy Laboratory. Golden, Colorado. Accessed June 27, 2008 at http://www.p2pays.org/ref/11/10516/refuse.html.

NREL/TP-570-25357. (1998). *Biomass Gasifier "Tars": Their Nature, Formation, and Conversion.* National Renewable Energy Laboratory. Golden, Colorado.

O'Leary, P. & Walsh, P. (2002). Landfill Continuing Education Course. Waste Age magazine, Penton Media, Inc. New York, New York. and University of Wisconsin-Madison, Solid and Hazardous Waste Education Center. Madison, Wisconson. Accessed at http://landfill-ed.wasteage.com.

PB-293 165. (1979). *Recovery, Processing and Utilization of Gas from Sanitary Landfills.* In: R., Ham, K., Hekimian, S., Katten, W. Lockman, & R. Lofy. *Lockman and Associates.* Monterey Park, California.

Phyllis, Composition of Biomass and Waste Database. Energy research Centre of the Netherlands. Petten, Holland, Netherlands. Accessed April 24, 2008 at http://www.ecn.nl/phyllis.

Repa, E. (2005). *NSWMA 's 2005 Tip Fee Survey.* National Solid Wastes Management Association. Washington, D.C.

Rezaiyan, J. & Cherenmisinoff, N. P. (2005). *Gasification Technologies A primer for Engineers and Scientists.* CRC Press. Boca Raton, Florida.

Simmons, P., Goldstein, G., Kaufman, S., Themelis, N. & Thompson, J. (2006). *The State of Garbage in America.* BioCycle., *47(4)*, 26.

Simmons, P., Goldstein, G., Kaufman, S., Themelis, N. & Thompson, J. (2006a). *The State of Garbage in America Recycling Data Analysis.* BioCycle, *47(10)*, 21.

Stevens, D. (1994). *Review and Analysis of the 1980-1989 Biomass*

Thermochemical Conversion Program. NREL/TP-42 1-7501. National Renewable Energy Laboratory. Golden, Colorado.

Stiles, D., Jones, S., Orth, R., Saffell, B., Stevens, D. & Zhu, Y. (2008). *Biofuels in Oregon and Washington: A Business Case Analysis of Opportunities and Challenges.* PNNL-17351. Pacific Northwest National Laboratory, Richland, Washington.

SWIS. (2008). Solid Waste Information System. California Integrated Waste Management Board. Sacramento, California. Accessed May 5, 2008 at http://www.ciwmb.ca.gov/SWIS/Downloads/SWIS.xls.

Themelis, N., Kim, Y. & Brady, M. (2002). *Energy recovery from New York City solid wastes.* ISWA Journal: Waste Management and Research 20(2002). 223-233.

Thermoselect. (2008). *Projects in Japan.* THERMOSELECT S.A. Locarno, Switzerland. Accessed September 26,2008 at *http://www.thermoselect. com/index.cfm?fuseaction=spezific&m=3.*

Westinghouse Plasma Corporation. 2007. *Projects Overview.* Madison, Pennsylvania. Accessed September 18, 2008 at http://www. westinghouse-plasma.com/projects/overview.php.

Westinghouse Plasma Corporation. (2007a). *Plasma Gasification Vitrification Reactor.* Madison, Pennsylvania. Accessed September 18, 2008 at http://www.westinghouseplasma.com/markets_applications /energy.

End Notes

[1] Conditionally exempt, small quantity hazardous waste is defined by the EPA as 220 pounds (or less) of hazardous wastes per calendar month, or 2.2 pounds (or less) of acute hazardous wastes per calendar month, or 220 pounds (or less) of spill cleanup debris containing hazardous waste per calendar month.

[2] MSW density is highly dependent upon the level of compaction. An average MSW density of 0.25 tons per cubic yard was assumed for this report.

[3] Disposal mass data was collected by searching publicly available internet databases and contacting landfill sites via telephone.

[4] Landfill site-specific data was not available for Alaska, Kansas, Montana, Rhode Island, Wyoming, and Florida. It is likely that 3 counties in Alabama, 1 county in Kansas, 1 county in Rhode Island and 6 counties in Florida could support facilities processing 2755 or more short tons per day.

[5] Electric power requirements for treating auto shredder waste, fly ash, medical waste ,and municipal solid waste with varying levels of pre-treatment (leading to the wide range in values given) have been reported.

In: Using Municipal Solid Waste for Fuel   ISBN: 978-1-61209-512-7
Editor: Samantha M. Feller   © 2011 Nova Science Publishers, Inc.

Chapter 2

MUNICIPAL SOLID WASTE (MSW) TO LIQUID FUELS SYNTHESIS, VOLUME 2: A TECHNO-ECONOMIC EVALUATION OF THE PRODUCTION OF MIXED ALCOHOLS

United States Department of Energy

SUMMARY

Biomass is a renewable energy resource that can be converted into liquid fuel suitable for transportation applications and thus help meet the Energy Independence and Security Act renewable energy goals (U.S. Congress 2007). However, biomass is not always available in sufficient quantity at a price compatible with fuels production. Municipal solid waste (MSW) on the other hand is readily available in large quantities in some communities and is considered a partially renewable feedstock. Furthermore, MSW may be available for little or no cost.

This report provides a techno-economic analysis of the production of mixed alcohols from MSW and compares it to the costs for a wood based plant. In this analysis, MSW is processed into refuse derived fuel (RDF) and then gasified in a plant co-located with a landfill. The resulting syngas is then catalytically converted to mixed alcohols. At a scale of 2000 metric tons per day of RDF, and using current technology, the minimum ethanol selling price at a 10% rate of return is approximately $1.85/gallon ethanol (early 2008 $). However, favorable economics are dependent upon the toxicity characteristics of the waste streams and that a market exists for the by-product scrap metal recovered from the RDF process.

ACKNOWLEDGMENT

The authors thank DOE's biomass program for funding this work and acknowledge the modeling work performed by the National Renewable Energy Laboratory (NREL), which serves as the basis for the gasification and syngas conditioning portion of the models.

1.0. INTRODUCTION

Biomass is a renewable domestic resource that has the potential to make a significant impact on domestic fuel supplies. However, due to the disperse nature of biomass, its cost rises with the quantity collected. Thus capital cost economies of scale can be difficult to achieve. Municipal solid waste (MSW) is an important biomass containing resource that remains largely untapped in the United States. Some landfills have installed landfill gas recovery systems or employ waste-to-energy via combustion and/or refuse derived fuel production. However, these uses represent only a portion of the nation's waste and generally have low efficiency. Using MSW as a fuel feedstock takes advantage of existing collection infrastructure and extends landfill life. In some locations using this feedstock could take advantage of existing sorting infrastructure (i.e., Material Recover Facilities or MRFs), thus reducing costs. It also provides a domestic source of feedstock for fuel and captures energy value that is otherwise literally buried. Importantly, it does not compete for food or cultivatable land and may be available for low or even negative fees. According to the Energy Information Agency, on average, fifty-six percent of MSW can be classified as biogenic and therefore can be considered partially renewable (EIA 2007).

This report analyzes the use of MSW to produce mixed alcohols via gasification using existing technologies and compares those results to a biomass based system. The MSW plant is assumed to be co- located with a landfill. Sensitivities to fees, return on investment, and technology improvements are discussed. The process model and cost models used in this work are based upon previous analysis (Aden 2005, Phillips 2007, Zhu 2009) and employs similar methodology.

2.0. PROCESS DESIGN BASIS AND MODELING

A simplified block diagram for the MSW to ethanol process is shown in Figure 2-1. In this system, MSW is first separated to remove recyclables and shredded and milled to reduce its size. It is then dried prior to gasification to synthesis gas. Then syngas is sent to a tar reformer and a scrubber. Syngas free of tars and particulates is sent to a sulfur removal unit to remove sulfur compounds. Then clean syngas is sent to a steam reformer to convert methane to hydrogen and carbon monoxide and to adjust the H_2/CO ratio to that required by the mixed alcohols synthesis. The syngas is then compressed and sent to mixed alcohol synthesis. The product stream is cooled and the unconverted syngas and gaseous products are separated from the product liquid. The liquid product is dried in a molecular sieve and fractioned into ethanol and higher alcohols. Steam generated in the processes is collected and sent to the steam cycle for power generation. Process steam is extracted from the turbines for use in the gasifier, steam reformer and various process heaters. The entire process is assumed to be co-located with a landfill of suitable size. The reference biomass based process is the same as the MSW process, without the physical separation step.

The base case assumes existing technology for the gas cleanup section (tar cracking and steam reforming) and the mixed alcohols production (Spath 2005, Aden 2005, Zhu 2009). The future 2012 goal case (Phillips 2007) gas cleanup and alcohol production are included as sensitivity.

2.1. Process Design Basis

The gasifier feed rate is assumed to be 2000 metric tones/day of dry feed (2205 short tons per day). The MSW to ethanol process consists of nine main areas. Each area is described in the follow paragraphs.

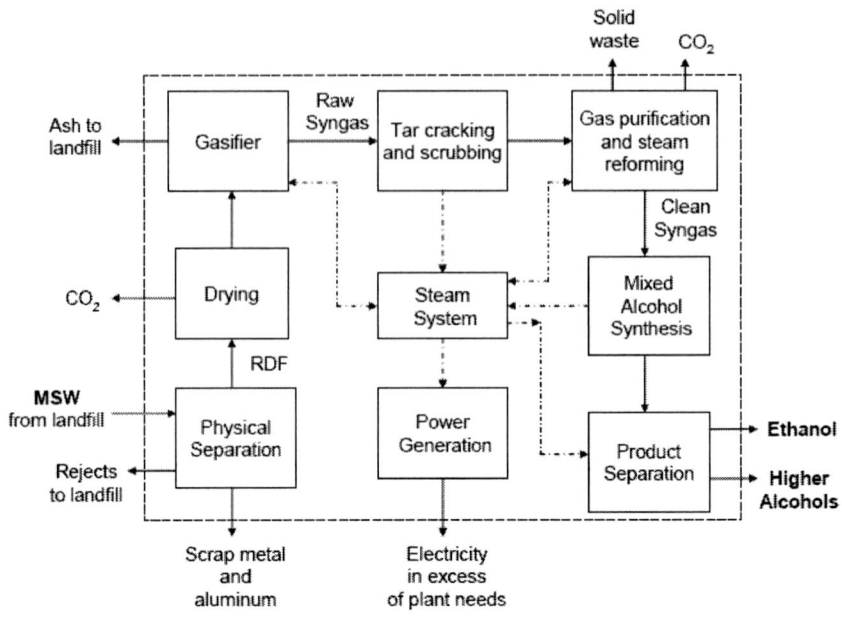

Figure 2-1. Process Diagram of the MSW to Ethanol Process.

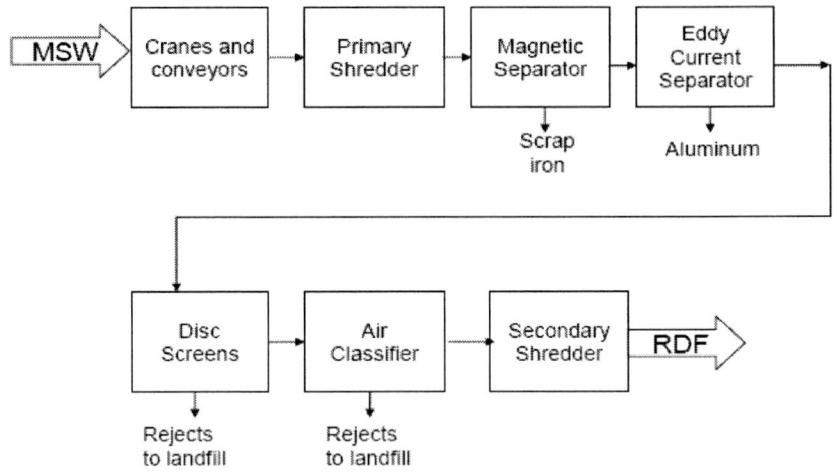

Figure 2-2. Process Diagram of the MSW to Ethanol Process.

2.1.1. Physical Separation

Raw MSW contains a large amount of non-combustible material, and therefore requires pre-processing before sending it to a gasifier. The pre-processing must be able to meet the requirements of the gasifier and be flexible enough to handle MSW variability. This flexibility must be in terms of the type of material handled and its frequency of delivery.

The pre-processing area is assumed to be similar to a Refuse Derived Fuel (RDF) facility. Some recyclables and non-combustibles are removed from the MSW to make a higher heating value product that is sized appropriately for gasification. The RDF separation equipment is located in an enclosed area. The building contains front end loaders, stockpiling areas, cranes, shredders, a ferrous magnet to remove scrap iron, an eddy current separator to capture aluminum, screens to separate by size, air classifiers to separate by weight, a series of conveyor belts and product storage areas. A simplified block diagram is shown in Figure 2-2.

Reject material contains broken glass, dirt and material too small to collect for the gasifier as well as some combustible material. Reject material plus recovered scrap metal is assumed to be 20% of the MSW feed rate. The rejects are returned to the landfill, while the scrap metal is stockpiled for sale. The RDF material retains significant mineral matter, resulting greater than 10% ash content.

The RDF facility is expected to operate on a schedule similar to that of the waste collection. For large municipalities, this could mean a twenty-four hour, seven days per week operation. Smaller facilities may only collect waste five days per week. Since the gasifier operates continuously at a feed rate of 2205 dry short tons per day (TPD), the RDF facility must run at a higher rate if it operates at less than round the clock. Table 2-1 shows the MSW feed rates required for various combinations of shifts per day and loss to meet 2205 dry short tons per day of RDF on a continuous basis. The term "loss" refers to the recovered metals and the reject material in the MSW that is not recovered in the RDF and is sold as scrap metal or returned to the landfill. It is assumed that the RDF facility in this study operates six days per week on two eight hour shifts per day at 20% loss, thus requiring an MSW feed rate of 200 short tons per hour (TPH). Two trains of 100 TPH MSW will allow flexibility to handle variations in MSW rates and equipment maintenance.

2.1.2. Drying

The feedstock (either RDF or biomass) at a moisture content of 50 wt% is fed at a rate of 2205 dry TPD. The biomass is assumed to be delivered at the correct size for gasification. The wet feed is dried in rotary dryers to a moisture content of 12 wt%. The dried feed is then conveyed to the gasifier.

Table 2-1. Example Operating Schedules for RDF Production

MSW Tons/hr	Shifts /day	Hours /shift	MSW Tons /day	Operating days /week	MSW Tons /week	MSW TPD continuous	MSW conversion to RDF		
							75%	80%	85%
143	3	8	3432	6	20,592	2942	**2206**	2353	2500
134	3	8	3216	6	19,296	2757	2067	2205	2343
126	3	8	3024	6	18,144	2592	1944	2074	**2203**
214	2	8	3424	6	20,544	2935	**2201**	2348	2495
201	2	8	3216	6	19,296	2757	2067	2205	2343
189	2	8	3024	6	18,144	2592	1944	2074	**2203**
429	1	8	3432	6	20,592	2942	**2206**	2353	2500
402	1	8	3216	6	19,296	2757	2067	2205	2343
378	1	8	3024	6	18,144	2592	1944	2074	**2203**

Note the higher ash content and lower heating value of RDF as compared to the wood. The ash from the RDF gasifier is assumed to be non-hazardous, but will likely require testing in the same way that ash from MSW incineration must be tested. Non-hazardous ash from incinerators can be returned to the landfill or used in roads and parking lots, depending upon local restrictions. (EPA 2008) However, some states regulate ash based on scale. For example, Washington State requires ash deposition to an ash monofill by plants processing more than 12 TPD of MSW (WAC 173-306-200).

2.1.3. Gasification

The indirectly-heated gasifier contains both a gasifier and a combustor. Dried feedstock is fed into a low- pressure indirectly heated gasifier. Steam extracted from the steam cycle is sent to the gasifier to fluidize the bed and to supply a portion of the heat required for the gasifier. The gasifier is mainly heated by circulating olivine particles between the gasifier and the separate combustor. Char and ash formed in the gasifier is carried out of the gasifier along with the olivine, separated in a series of cyclones and sent to the fluidized bed combustor, where air is used to burn the char, thereby reheating the olivine. The heat balance in the gasifier is achieved by adjusting air to the combustor, and the olivine circulation rate.

Table 2-2. RDF and Biomass Feedstock Quality and Gasifier Conditions

	RDF	Biomass (Poplar)
Feedstock analysis, % dry basis		
volatile matter	79.6	83.8
fixed carbon	10.0	15.3
ash	10.5	0.92
carbon	45.5	51.0
hydrogen	5.8	6.1
nitrogen	0.3	0.2
sulfur	0.2	0.1
oxygen (by difference)	37.8	42.3
BTU/lb, dry basis	7621	8671
Moisture Content, % wet	50%	50%
Gasifier temp, °C (°F)	822 (1511)	870 (1598)
Feed moisture, %	12	12
Gasifier press, psig	8	8
Steam rate, lb/lb dry feedstock	0.398	0.398
Combustor temp, °C (°F)	943 (1730)	995 (1823)
Combustor press, psig	8	8
Dry syngas composition, vol%		
hydrogen	17.6	24.0
carbon monoxide	38.8	42.4
carbon dioxide	9.3	12.8
methane	15.6	15.4
ethane	1.0	0.3
ethylene	16.7	4.4
acetylene	0.4	0.4
benzene	0.3	0.1
naphthalene	0.5	0.2
H2/CO ratio	0.45	0.57
Product gas HHV, btu/scf (dry)	663	468

    The indirectly-heated gasifier is modeled using the correlations reported in Spath, *et al.* (2005). The correlations are based on data from the Battelle-Columbus Laboratory (BCL) process development unit (PDU) gasifier. The RDF data are from the BCL PDU running on RDF as reported in Paisley, *et al.* (1990). These data are summarized in Table 2-2. Economics for both RDF and Biomass are included in this report.

2.1.4. Tar reforming and gas scrubbing

During gasification, a relatively small fraction of the feedstock is converted into tars consisting mostly of aromatic and poly-aromatic type hydrocarbons. The raw gas from the cyclone in the gasifier section is sent to a catalytic tar cracker, which is assumed to be a bubbling fluidized bed reactor. A portion of the tar, methane, and other light hydrocarbons in the raw gas are converted to CO and H_2. The gasifier also produces a small amount of NH_3 from the nitrogen in the biomass, which is then converted in the tar cracker to N_2 and H_2. The conversion percentage for each compound is reported in Spath, et al., (2005) and represents the state of technology. The gas enters the tar reformer at the gasifier outlet temperature and exits the reformer at 751°C (1,383°F). The syngas is further cooled to 150°C (300°F) and sent to a wet scrubber to remove other impurities, such as particulates, NH_3, and some residual tars.

RDF is expected to contain variable amounts of chlorides depending upon the amount of plastics such as polyvinyl chloride in the MSW stream. Halides are assumed to be removed in the particulate scrubber by the addition of lime.

2.1.5. Gas purification and steam reforming

The scrubbed syngas is compressed to 450 psia in preparation for gas purification. Mercury and sulfur are the main contaminants that must be removed prior to stream reforming.

The RDF material potentially contains mercury which becomes volatile in the gasifier (Parsons 2002). Mercury in MSW mainly comes from fluorescent bulbs and its concentration can be as high 6000 ppb (EPA 1997). Carbon beds are an effective means of mercury removal from syngas (Parsons 2002) and a series of fixed bed carbon vessels are assumed. The spent carbon containing adsorbed mercury is disposed of as hazardous waste.

A liquid phase oxidation (LO-CAT) process followed by a ZnO bed is used to remove sulfur. The LO- CAT process is assumed to remove the sulfur to a concentration of 10 ppm H_2S, and then the ZnO bed polishes the syngas to less than 1 ppmv (Spath, et al. 2005).

Syngas leaving the ZnO bed is sent to a steam reformer to convert the remaining methane and light hydrocarbons to additional syngas and to adjust the H_2:CO ratio via the water-gas shift reaction. The main steam reforming reactions are:

$$C_nH_m + nH_2O \leftrightarrow (n+m/2)H_2 + nCO \qquad (1)$$

$$CO + H_2O \leftrightarrow CO_2 + H_2 \qquad\qquad (2)$$

Before the syngas is sent to the steam reformer, it is mixed with high temperature steam and compressed carbon dioxide (from the amine system as discussed below). Reactions take place between 800 and 900°C (1472 and 1652°F). The H_2: CO ratio is adjusted to approximately 1.2, as required by the mixed alcohol synthesis reaction. The converted syngas passes through several heat exchangers to recover heat by generating saturate high pressure steam and superheated high pressure steam. The cooled syngas from reforming process is further cooled by air cooling and cooling water. The cooled syngas is further sent to an amine unit to remove most of the CO_2 which is a diluent in the high pressure synthesis system. The clean syngas is compressed to 2000 psi and sent to the mixed alcohol synthesis section. The steam reformer is fired with off-gas from the mixed alcohol synthesis.

2.1.6. Mixed alcohol synthesis

The mixed alcohol synthesis involves multiple reactions with different pathways to various alcohols and hydrocarbons. The overall stoichiometric reaction for higher alcohol synthesis is:

$$nCO + 2nH_2 \rightarrow CnH_{2n+1}OH + (n-1)H_2O \qquad\qquad (3)$$

where the value of "n" typically ranges from 1 to 6. Hydrocarbon synthesis takes place according to a similar reaction scheme. While the stoichiometry of these reactions suggests an optimum H_2/CO ratio approximately 2, the optimal ratio is closer to 1.0 if the catalyst is significantly active for the water-gas shift reaction. This study assumes a modified Fischer-Tropsch catalyst (K/Co/MoS catalyst) that reflects the current state of technology, and thus represents a scenario that might be obtainable today. A description of this catalyst can be found in Aden et al, (2005).

Clean syngas at 2000 psi is preheated to 299°C (570°F) in a feed-product exchanger. The mixed alcohol reactor is assumed to be of a fixed bed tubular design, with catalyst in the tubes and steam raised in the shell. The product gas is partially cooled against the inlet compressed syngas, followed by further cooling to condense the alcohols and water. Most of unconverted syngas is recycled to the stream reformer. The methanol product is recycled back to the mixed alcohol synthesis reactor to increase the conversion efficiency. The liquid alcohols are then sent to the alcohol separation and purification processes. Methanol purge and product gas purge streams are combined and sent to the fuel gas system. The assumed mixed alcohol synthesis reaction conditions and per pass conversions for are shown in Table 2-3. The specific conversions of CO in each of the main reactions are set in order to reach catalyst performance targets consistent with those of Aden, et al. (2005).

The large heat release from the exothermic mixed alcohol reactor is removed by vaporizing boiler feed water on the shell side of the reactor. The high pressure steam is sent to the steam cycle and power recovery section for electric power generation.

Table 2-3. Mixed Alcohol Synthesis Assumptions

Parameter	Values
Temperature (°F)	570
Pressure (psia)	2000
H_2/CO Ratio	1.2
CO_2 inlet concentration	0.2 mol%
Gas hourly space velocity, v/h/v	3000
$CO + H_2$ Reactions	**Mole % CO Conversion per pass**
$CO + H_2O \rightarrow CO_2 + H_2$	13%
$CO + 3 H_2 \rightarrow CH_4 + H_2O$	4.5%
$2 CO + 4 H_2 \rightarrow C_2H_6 + H_2O$	0.5%
$CO + H_2 \rightarrow$ Methanol	4.1%
$2 CO + 4 H_2 \rightarrow$ Ethanol + H_2O	11.4%
$3 CO + 6 H_2 \rightarrow$ Propanol + $2 H_2O$	3%
$4 CO + 8 H_2 \rightarrow$ n-Butanol + $3 H_2O$	1%
$5 CO + 10 H_2 \rightarrow$ n-Pentanol + $4H_2O$	0.5%
Methanol Recycle Reactions	**Mole % Recycled Methanol Conversion**
Methanol + CO + $2 H_2 \rightarrow$ Ethanol + H_2O	58%
Methanol + 2 CO + $4H_2 \rightarrow$ Propanol + $2 H_2O$	7%
Methanol + 3 CO + $6 H_2 \rightarrow$ n-Butanol + $3 H_2O$	4.5%
Methanol + 4 CO + $8 H_2 \rightarrow$ n-Pentanol + $4H_2O$	2%

2.1.7. Product separation and purification

The raw mixed alcohol product from the synthesis step is dried with a molecular sieve. The dried product is then distilled into a methanol stream that is recycled to the synthesis reactor, a purified ethanol stream and a higher alcohol stream. The higher alcohol product contains propanol and higher boiling alcohols.

2.1.8. Power generation

Saturated steam is generated by cooling the process streams in the gasifier, steam reformer and mixed alcohol synthesis areas. Saturated steam is superheated in the steam reformer section then sent to a steam turbine to generate power for the plant and to provide process steam for use in the system.

Table 3-1. Process Model Assumptions

Parameter	Biomass Case	MSW Case
% MSW in RDF	Not applicable	80%
Dryer		
Feed inlet moisture, wt%	50	50
Outlet moisture, wt%	12	12
Gasifier		
Pressure, psi	23	23
Temperature, °C (°F)	822 (1511)	870 (1598)
Bone dry feed, metric ton/d	2000	2000
Tar Reformer, T, °C (°F)/ P, psi	751 (1383) / 23	751 (1383) / 23
Steam Reforming		
Temperature, °C (°F)	900 (1652)	900 (1652)
Pressure, psia	435	435
H_2:CO in reformed syngas	2.1	2.1
Mixed alcohol Synthesis and Purification		
Temperature, °C (°F)	299 (570)	299 (570)
Pressure, psia	2000	2000
Methanol Recycle, %	90	90
Steam System		
Pressure, psia	800	800
Superheat temperature, °C (°F)	538 (1000)	538 (1000)

2.2. Analysis Approach

The process simulation was developed in CHEMCAD and the capital and operating costs were assembled in an EXCEL spreadsheet. A discounted cash flow analysis is used to estimate the ethanol selling price.

3.0. SIMULATION AND ECONOMIC ASSUMPTIONS

The main assumptions for the performance and cost models are described in this section.

3.1. Simulation Assumptions

Table 3-1 shows the main assumptions for the biomass and MSW simulations. The main difference is in the composition of the feedstocks.

3.2. Economic Assumptions

Figure 3-1 plots the total capital investment required for an RDF facility as taken from various literature sources (source dates are listed in the legend) and converted to January 2008 dollars using the Chemical Engineering Index (CEI 2008). Included in each sources estimate are all direct and indirect capital costs to provide sufficient equipment to produce RDF from MSW and recover recyclables. The estimates also include scales, a processing building and all site work. Note that the trend of the capital investment increase is almost linear with the increase in plant scale. RDF facilities typically have multiple lines to accommodate equipment maintenance and variable processing rates and therefore have little economy of scale.

The capital investment for the rest of the mixed alcohols plant is determined from a cost rollup of the specific equipment needed for each area of the plant (e.g. dryer, gasifier, heat recovery, mixed alcohol reactor). Most of the base equipment costs for the gas purification and conditioning and the steam cycle and power generation sections of the plant come from Spath, *et al.* (2005). The estimation of the equipment costs for the indirectly heated gasifier is based on Hamelinck and Faaij (2002). The equipment costs for mercury removal is from Parsons (2002). The mixed alcohol synthesis and product separation equipment is sized using the heat and material balances, and the associated equipment costs come from APSEN ICARUS.

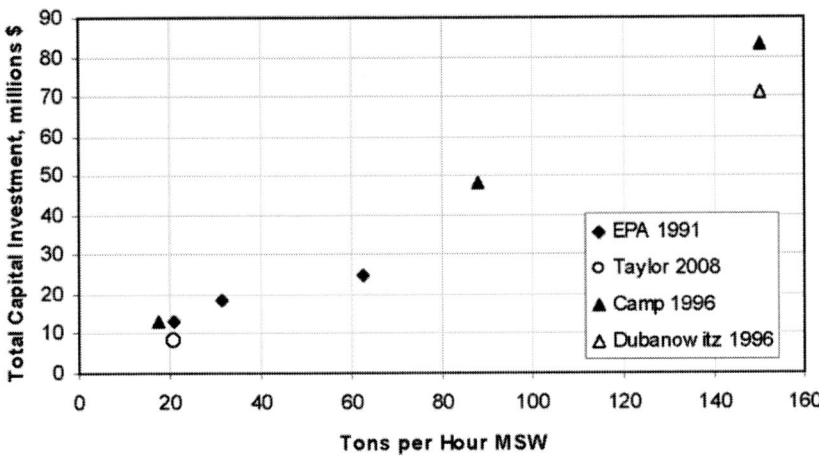

Figure 3-1. RDF Facility Scale vs. Total Capital Investment.

Table 3-2. Total Project Investment Assumptions

Total Purchased Equipment Cost for Mixed Alcohols Portion of the Plant (TPEC)	100% of TPEC
Purchased Equipment Installation	39%
Instrumentation and Controls	26%
Piping	31%
Electrical Systems	10%
Buildings (including services)	29%
Yard Improvements	12%
Total Installed Cost (TIC)	247%
Indirect Costs	
Engineering	32%
Construction	34%
Legal and Contractors Fees	23%
Project Contingency	37%
Total Indirect	126%
Capital Investment for the Mixed Alcohols Portion of the Plant	373%

Table 3-3. Economic Assumptions

	Value used in model, 2008 basis	Units or Basis	Reference
Raw Materials			
Hybrid poplar chips	60	$/dry short ton	Aden, 2008
Olivine makeup	207	$/short ton	Phillips, *et al.* 2007*
Ash disposal	34	$/short ton	Valkenburg, *et al.* 2008
Tar cracker catalyst	6.15	$/lb	Phillips, *et al.*, 2007*
Reformer catalyst	21.9	$/lb	SRI PEP 2007*
Mixed alcohol catalyst	5.25	$/lb	Phillips, *et al.* 2007
Carbon	8.34	$/1b	Parsons, 2002*
By-Products			
Scrap metal	0.2	$/lb	USFN 2008
Scrap aluminum	1.0	$/lb	Thompson 2008
Higher alcohol value	1.15	$/gallon	Phillips, *et al.* 2007
Sulfur	40	$/ton	Phillips, *et al.* 2007*
Waste By-Products			
Waste water treatment	2.47	$/100 ft^3	Phillips, *et al.* 2007*
Hazardous waste disposal	500	$/ton	Parsons, 2002
Non-hazardous waste disposal	34	$/ton	Valkenburg, *et al.* 2008
Utilities			
Cooling water	168	¢/1000 gal	Phillips, *et al.* 2007*
Electricity	6.27	¢/kWh	EIA 2008, industrial price
Stream Factor	90%		estimated
MACRS Depreciation, yrs	7		Phillips, *et al.* 2007
Plant life, yrs	20		Phillips, *et al.* 2007
Construction Period 1st 6 months expenditure Next 12 months expenditure Last 12 months expenditure	2.5 years 8% 60% 32%		Phillips, *et al.* 2007
Start-up time Revenues Variable Costs Fixed Costs	6 months 50% 75% 100%		Phillips, *et al.* 2007
Working Capital	5% of Total Capital Investment		Phillips, *et al.* 2007
Land	6% of Total Purchased Equipment Cost (taken as 1st year construction expense)		Phillips, *et al.* 2007
Internal Rate of Return	10%		Phillips, *et al.* 2007
* Reference value escalated to early 2008 dollars using the producer price index			

All capital costs are reported in January 2008 dollars. The total capital investment for the mixed alcohol plant excluding the RDF facility is factored from installed equipment costs using the factors shown in Table 3-2. The capital investment for the RDF facility is added to the capital investment for the remainder of the plant to determine the total capital required.

Table 3-3 lists the assumptions used to estimate the production costs. The minimum ethanol selling price (MESP) was determined using a discounted cash flow rate of return analysis similar to that used in Phillips *et al.* 2007. The MESP is the selling price of the fuel that makes the net present value of the process equal to zero for a specified discounted cash flow rate of return over a 20 year plant life. A sensitivity analysis was conducted to determine the effect of different financial and operating assumptions on the MESP.

4.0. RESULTS AND ANALYSIS

This section describes the main performance and cost simulation results for each scenario.

4.1. Performance Results

Table 4-1 shows the main performance results for the MSW case and compares it to the same process fed only with biomass. Both cases produce higher alcohols and sulfur as by-products. The mixed alcohol yield is lower for the MSW case due to the higher fraction of inorganic material in the RDF than in the wood. Both processes also generate electricity in excess of the plant's needs. Additionally, the MSW based process recovers metals that can be sold as scrap. On the other hand, the MSW case produces more waste streams, although one could argue that the wastes are a fraction of the total MSW waste to begin with, and thus a net gain to the landfill in terms of landfill usage. The non-hazardous MSW waste stream consists of ash, MSW in the materials separation pre-processing area that is not recovered metals or RDF, and spent olivine. However, the spent carbon beds contain mercury and therefore must be disposed of as hazardous waste.

Table 4-1. Main Performance Analysis Results

Case	MSW Case	Biomass Case
Feedstock		
Wood chips or MSW, dry million lb/y	1930 MSW; 1,447 RDF	1,447 Wood chips
Products		
Ethanol, mmgal/y	27	36
Propanol and high alcohols, million gal/y	9	12
Sulfur, lb/y	81,600	35,200
Recyclable scrap aluminum, million lb/y	110	0
Recyclable scrap iron, million lb/y	25	0
Yields		
Ethanol, gal/ton dry feedstock	28 MSW basis; 38 RDF basis	50
Higher alcohols, gal/ton dry feedstock	9 MSW basis; 13 RDF basis	17
Waste Products		
Hazardous waste, lb/y	80,000	0
Total Non-hazardous solid waste, million lb/y	390	22.5
Ash, million lb/y	152	19
Spent olivine, million lb/y	3	3.5
MSW rejects, million lb/y	235	0
Net Power for export, MW	9.3	3.2

Table 4-2. Capital Costs for the MSW and Biomass Cases

	MSW Case		Biomass Case	
Million gallons/year ethanol	27		36	
Capital Costs	Million $	% of Total	Million $	% of Total
RDF production	$105	23%	$0	
Feedstock drying	$40	9%	$39	11%
Gasification, tar reforming, scrubbing	$67	15%	$56	16%
Syngas conditioning	$164	37%	$167	49%
Mixed alcohol synthesis	$20	4%	$21	6%
Mixed alcohol separation	$9	2%	$11	3%
Steam system and power generation	$34	8%	$39	11%
Remainder off-site battery limits	$9	2%	$9	3%
Total Capital Investment	$449		$343	
Project investment/annual gallon ethanol	16		9	

4.2. ECONOMIC RESULTS

Table 4-2 shows the capital cost breakdown for each section of the plant. The RDF facility amounts to almost a quarter of the required investment. Gasification and syngas cleanup make up the bulk of the rest of the costs. Improved and simplified syngas cleanup and improved mixed alcohols synthesis is an ongoing area of research. It is expected that costs for these areas will decrease with time. Reduction in capital and improved yields are addressed in the sensitivity section.

Table 4-3 shows the operating cost breakdown for both cases. The MSW case assumes that the plant is co-located with a municipal waste landfill and that the MSW feedstock is free. This case also assumes that the non-hazardous waste is returned to the landfill for no fee. The MSW case has lower variable costs than the biomass case due to no feedstock cost and additional by-product credits from the sale of the recyclables. The fixed costs for the MSW case are higher, mainly due to the larger work force needed to produce the RDF material. The gasification and mixed alcohol production areas are assumed to be highly automated and require fewer operators per unit than does the labor intensive RDF facility. Waste treatment costs are relatively low for both cases. Disposal of the mercury laden spent carbon is not a large cost factor for the MSW case. The higher capital costs for the MSW case are more than offset by the lower operating costs. The base MSW case selling price of $1.85 per gallon of ethanol falls within the selling price of ethanol over the past year. Ethanol prices from June 2007 to October 2008 have fluctuated up and down between $1.60 and $2.50/gallon (ICIS 2008). However, a 10% internal rate of return on investment may not be sufficiently high to make this economically attractive to investors. Sensitivities to return on capital are explored in the next section.

4.3. Sensitivity Analysis

A sensitivity analysis was conducted to investigate the effects of different cost assumptions such as return on investment, MSW fee, and by-product credits.

Table 4-3. Economic Results for the MSW and Biomass Cases

	MSW Case	Biomass Case
Million gallons/year ethanol	27	36
Operating Costs	$/gal	$/gal
Raw materials		
Feedstock (MSW or Biomass)	0.00	1.19
Catalysts & Chemicals	0.10	0.09
By-product credits		
Higher alcohols	-0.39	-0.39
Scrap Aluminum	-0.86	0.00
Scrap Iron	-0.58	0.00
Electricity sold to grid	-0.17	-0.04
Waste treatment or Disposal		
Gasifier ash	0.000	0.01
MSW rejects	0.000	0.00
Spent carbon	0.000 1	0.00
Waste water treatment	0.026	0.02
Total variable cost, $/gal ethanol	-1.86	0.87
Fixed costs, $/gal ethanol	0.87	0.51
Capital depreciation, $/gal ethanol	0.82	0.47
Average income tax, $/gal ethanol	0.56	0.33
Average return on investment (10% IRR)	1.46	0.86
Estimated Selling Price (10% IRR), $/gal ethanol	1.85	3.05

4.3.1. Effect of MSW Fee and Return on Investment

According to a 2004 survey, as reported in Valkenburg *et al.* 2008, tipping fees vary widely across the country, from a low of $24.06 per ton in the south to $70.06 per ton in the Northeast, with the average of approximately $34.29/ton nationwide. Tipping fees for MSW that was incinerated averaged about $61.64 nationwide. Thus there is precedent for increasing tipping fees to offset capital costs, assuming the local market will bear the cost increase. Figure 4-1 shows how varying the tipping fee affects the minimum ethanol selling price at a 10% IRR.

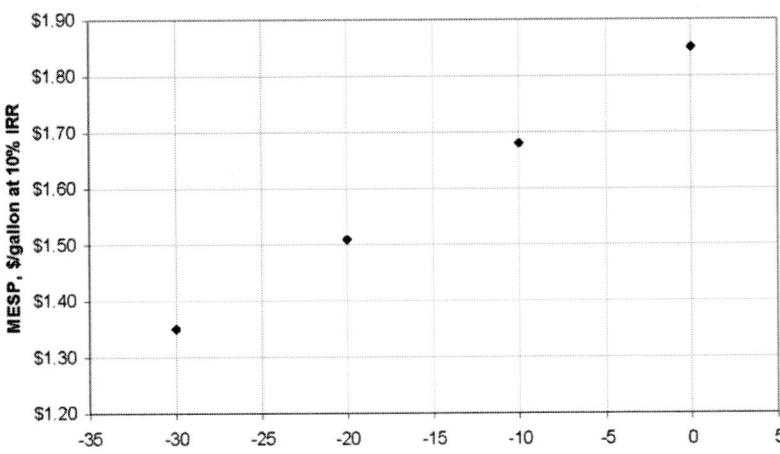

Figure 4-1. Effect of Tipping Fee on MESP.

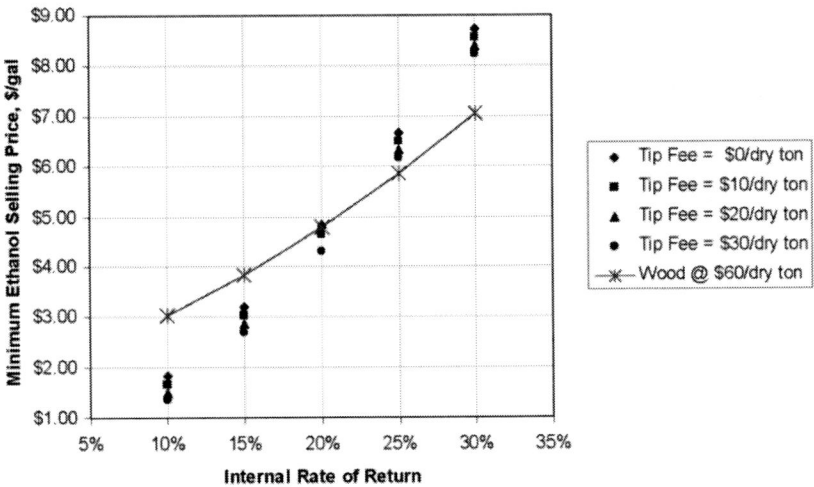

Figure 4-2. Effect of Tipping Fee on MESP at Different Rates of Return.

Recall that the tipping fee is the cost charged by the landfill to waste generators; a tipping fee of $30/ton is the same as a feedstock cost of -$30/ton. Increasing the tipping fee from $0 to $30 per ton is worth approximately $0.50/gallon of ethanol.

Figure 4-2 expands Figure 4-1 to show the effects of increasing the return rate. This plot shows several things:

- Even with a negative feedstock cost, the MSW case can at best return about 10% of the capital before the selling price of ethanol exceeds $2.00/gallon.
- As shown previously in Table 4-2, the RDF facility significantly increases the capital cost of the MSW plant compared to the biomass (wood) plant. The MSW case and the biomass (wood) case MESP overlap when the IRR is approximately 20%. This is mainly due to the tradeoff between higher capital and lower feedstock cost for the MSW case as compared with the lower capital and higher feedstock costs for the biomass case. Above an IRR of 20%, the higher capital costs for the MSW case offset the gains made by low or no feedstock cost.
- There do not appear to be incentives to decrease the scale of the plant from 2205 TPD RDF, as this is likely to increase production costs and thus lower the return on investment.

4.3.2. Effect of RDF Operating Schedule and MSW and By-Product Value

Figure 4-3 shows the effect of selected operating variables. They are shown as cost differences with the base case assumptions of 2205 TPD MSW, 10% IRR, and no cost for the MSW. The top two bars show the sensitivity to RDF processing schedules. The base case assumes that the RDF plant processes MSW on two eight-hour shifts/day for six days per week, while the gasifier and alcohol synthesis systems operate continuously. The base case also assumes that 80% of the MSW is recovered as RDF. Forty cents per gallon could be saved by continuous 24 hour per day, seven day per week RDF processing. This is mostly due to the smaller capital investment for the RDF plant. Maintaining the same operating schedule as the base case, but with only 75% recovery of MSW as RDF increases costs by about twenty cents per gallon.

The middle three bars show the effect of fees. In the base case, the MSW is assumed to be free of charge and the waste generated in the RDF plant is landfilled at no additional cost. As noted previously, mercury in the MSW is assumed to be part of the RDF and captured post gasification in a carbon bed. The carbon bed is then disposed of as hazardous waste. If the reject MSW material (glass and fines) in the RDF plant contains significant amounts of mercury, say from broken fluorescent bulbs, then that material might be reclassified as hazardous. Disposal of the MSW reject material as hazardous waste increases the ethanol selling price by over two dollars per gallon. While this may not be a likely scenario, it does illustrate the need to sample and test waste materials for proper disposal.

If the facility is not owned by the landfill, but still adjacent to it, then the landfill may charge a fee for providing the MSW and fees to return the rejected solids. This is just the opposite of the tipping fee case presented in the previous section. In the tipping fee case, the landfill charges the waste generators to leave the MSW, and thus the feedstock itself is potentially income generating. This time, rather than a negative cost feedstock, the mixed alcohol plant must purchase the MSW from the landfill. Assuming that +$34/ton is charged for the MSW to the RDF plant and another +$34/ton is charge to return the waste generated in RDF production, then the cost of ethanol rises by about $1.40/gallon. If the landfill does not charge a fee for the MSW, but does require payment for ash rejects, then the selling price increases by about twenty cents per gallon (as shown in the bottom bar)

The lowest two bars show the effects of by-products on MESP. Reclaiming and selling scrap iron and aluminum (noted as recyclables on the chart) is important for process economics. Over one dollar per gallon is lost if the recyclables have no value (again assuming that the MSW is obtained at no cost to the RDF plant).

4.3.3. Effect of Technology Improvements

The conservative mixed alcohol case represents a scenario that might be possible to implement today. However, as shown in Figure 4-2, the economics are attractive only at about 10% IRR. Technology improvements can help reduce the capital cost and improve the overall economics. Thus the goal case thermo-chemical mixed alcohol process as described in Phillips et al. (Phillips 2007) was analyzed assuming MSW with RDF processing. As described in that report, improvements to the tar reformer, elimination of the steam reformer, simplified syngas conditioning and improved mixed alcohol catalyst yields reduce the capital and operating costs. A comparison between the main assumption differences and main results for the conservative MSW case and the goal MSW case are shown in Table 4-4. Assumptions not shown are the same as those listed in Table 3-1.

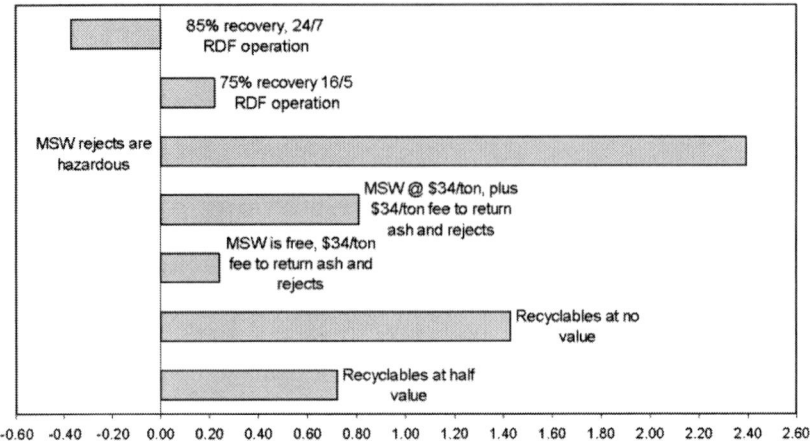

Figure 4-3. MESP Sensitivity to Process and Operating Changes.

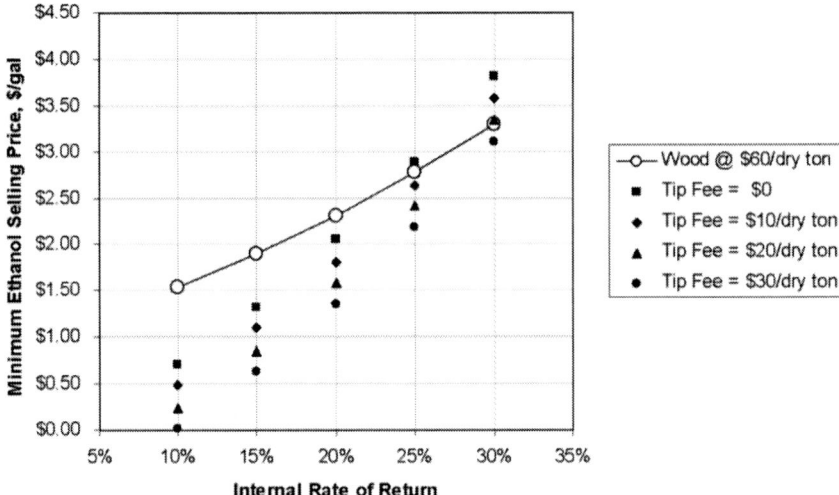

Figure 4-4. Effect of Improved Technology.

As can be seen in Figure 4-4, improved economics allow selling prices of less than $2/gallon without a tipping fee for a 20% rate of return if these improvements were available today. Higher returns on investment can be had by increasing the tipping fee. The wood case represents the goal case at a feedstock cost of $60/ton.

Table 4-4. MSW Conservative and MSW Goal Cases Comparison

Assumption Differences	MSW Conservative Case	MSW Goal Case
Tar Cracker/Reformer, T, °C (°F) / P, psi	751 (1383) / 23	889 (1633) /2 3
Steam Reforming	Yes	No
Steam System		
Pressure, psia	800	850
Superheat temperature, °C (°F)	538 (1000)	482 (900)
Mixed Alcohol Synthesis		
Temperature, °C (°F)	299 (570)	299 (570)
Pressure, psi	2000	900
Methanol Recycle, %	90%	97%
Liquid Hourly Space Velocity, v/v/h	3000	6000
Catalyst life, years	1	5
Feedstock Cost, $/short ton	0	0
Performance Results		
Products		
Ethanol, mmgal/y	27	42
Propanol and high alcohols, million gal/y	9	7
Yields		
Ethanol, gal/ton dry feedstock	28 MSW basis; 38 RDF basis	44 MSW basis; 55 RDF basis
Higher alcohols, gal/ton dry feedstock	9 MSW basis; 13 RDF basis	8 MSW basis; 12 RDF basis
Net power for export. MW	9	3
Cost Results		
Capital cost, millions $ (2008 basis)	$449	$313
Operating Costs, $/gal ethanol		
Raw materials	0.10	0.04
Waste disposal	0.03	0.00
By-product credits	-1.83	-1.03
Electricity	-0.17	-0.04
Fixed costs	0.87	0.44
Depreciation	0.82	0.37
Average income tax	0.56	0.25
Average return on investment	1.47	0.68
Minimum ethanol selling price, $/gal (10% IRR)	1.85	0.71

CONCLUSIONS AND RECOMMENDATIONS

A techno-economic analysis of an MSW to ethanol process via gasification was conducted. The base case assumes that the MSW is obtained at zero cost and that the metals are recovered and sold as a byproduct. The base case process is based on currently available technology for syngas cleanup and mixed alcohols synthesis. The results of this study provide information about the main performance and economics for the systems.

Sensitivity analysis was also conducted to investigate the effects of key assumptions on ethanol selling prices.

The capital required to produce mixed alcohols from MSW rather than from biomass is higher, due to the need to process the MSW into RDF. However, landfill-based operations offer the opportunity for zero or negative cost feedstock which offset the cost higher capital cost. The simulation results showed that the estimated ethanol selling price of $1.85/gallon (early 2008 dollars, 10% return on investment) for the indirectly-heated gasifier system, and assuming that the by-product metals are sold at a profit. $1.85/gallon is competitive with early 2008 ethanol market prices, thus this process might be implemented today using existing technology. Other conclusions are as follows:

- The sale of RDF plant recyclables are a necessary part of an economic operation.
- Returns on investment of greater than 10% may be achieved by technological improvements in the syngas cleanup and mixed alcohol synthesis steps. In particular, implementing the 2012 time- frame improvements outlined in Phillips et al. (Phillips 2008) reduce the MESP by half.
- Waste streams from the process need to be tested for toxicity.

In addition, further work is needed to clarify the effect of MSW variability on RDF quality and gasifier operation. It would be useful to determine the economics of more mature products from MSW generated syngas, in particular, methanol, Fischer-Tropsch fuels, gasoline, synthetic natural gas as well as alternate means of processing besides gasification.

MSW is a potentially valuable, partially renewable feedstock that has been under utilized. This study provides potential research areas for process improvement.

REFERENCES

Aden, A. (2008). *Biochemical Production of Ethanol from Corn Stover: 2007 State of Technology Model.* NREL/TP-5 10-43205. National Renewable Energy Laboratory, Golden, CO. May 2008. *http://www.nrel.gov/docs/fy08osti/43205.pdf.*

Aden, A., Spath, P. & Atherton, A. (2005). *The Potential of Thermochemical*

Ethanol Via Mixed Alcohols Production. Milestone Completion Report, FY05-684. National Renewable Energy Laboratory, Golden, CO. October.

Caputo, A. C. & Palegagge, P. M. (2002). RDF Production Plants: I Design and Costs. *Applied Thermal Engineering, 22*, 423-437.

Chang, N., Chang, Y. & Chen, W. C. (1997). Evaluation of Heat Value and Its Prediction for Refuse- Derived Fuel. *The Science of Total Environment*, 139-148.

Chemical Engineering Index. 2008. www.che.org.

Dubanowitz, A. (2000). *Design of a Materials Recovery Facility (MRF) for Processing the Recyclable Materials of New York City's Municipal Solid Waste*. MS Thesis, Columbia University. May.

EIA, (2007). *Methodology for Allocating Municipal Solid Waste to Biogenic and Non-Biogenic Energy*. DOE/EIA-0226. Energy Information Administration, Office of Coal, Nuclear, Electric and Alternative Fuels, U.S. Department of energy, Washington, DC. May. *http://www.eia.doe. gov/cneaf/solar.renewables/page/mswaste/ msw.pdf*.

EIA, (2008). *Monthly Energy Review Average Retail Prices of Electricity*. January 2008 Industrial US Average. Energy Information Administration, Office of Coal, Nuclear, Electric and Alternative Fuels, U.S. Department of energy, Washington, DC. *http://www.eia. doe.gov/emeu/mer/pdf/pages/ sec9_14.pdf*.

EPA (2008). *Municipal Solid Waste: Electricity from Municipal Solid Waste*. EPA website accessed 10/2008. *http://www.epa.gov/cleanenergy/ energy*.

EPA (1997). *Mercury Study Report to Congress, Volume 2*. EPA-452/R-97-004. December 1997. *http://www.epa.gov/ttncaaa1/t3/reports/ volume2. pdf*.

Hamelinck, C. N. & Faaij, A. P. C. (2002). Future Prospects for Production of Methanol and Hydrogen from Biomass. *Journal of Power Sources, 111(1)*, 1-22. 18 September 2002.

ICIS (2008). ICISPricing. Subscription pricing service available at: http://www.icispricing.com.

Niessen, W. R., Marks, C. H. & Sommerlad. R. E. (1996). *Evaluation of Gasification and Novel Thermal Processes for the Treatment of Municipal Solid Waste*. Camp, Dresser & McKee. National Renewable Energy Laboratory. NREL/TP-430-2 1612. August 1996. *http://www.osti.gov/ energycitations/servlets/purl/10164285-7CBnFx/webviewable/ 10164285.PDF* .

Paisley, M. A., Creamer, K. S., Tewksbury, T. L. & Taylor, D. R. (1989). *Gasification of Refuse Derived Fuel in the Battelle High Throughput*

Gasification System. PNL-6998. Pacific Northwest National Laboratory, Richland, WA.

Parsons, (2002). *The Cost of Mercury removal in an IGCC Plant. Final Report to DOE.* September 2002. Parsons Infrastructure and Technology Group Inc. *http://www.netl.doe.gov/technologies/coalpower/ gasification.*

Peer (1991). *Material Recovery Facilities for Municipal Solid Waste.* EPA 625/6-91/031. Peer Consulting and CalRecovery, Inc. United States Environmental Protection Agency. September 1991. *http://www.epa. gov/nrmrl/pubs/625691031/625691031.pdf.*

Phillips, S., Aden, A., Jechura, J. & Dayton, D. (2007). *Thermochemical Ethanol via Indirect Gasification and Mixed Alcohol Synthesis of Lignocellulosic Biomass.* NREL/TP-5 10-41168. April.

Spath, P., Aden, A., Eggerman, T., Ringer, M., Wallace, B. & Jechura, J. (2005). Biomass Hydrogen Production Detailed Design and Economics Utiltizing the Battelle Columbus Laboratory Indirectly Heated Gasifier. NREL/TP-5 10-37408. National Renewable Energy Laboratory, Golden, CO.

SRI PEP (2003). Yearbook International. SRI Consulting, United States. Menlo Park, CA. 2003. *SRI PEP 2007 Yearbook International.* SRI Consulting, United States. Menlo Park, CA. 2007.

Taylor, J. T. (2008). Taylor Recycling Facility Pre-Development of Montgomery Site for Biomass Energy Generation Final Report. Contract 8997. http://www.taylorbiomassenergy.com/Images/NYSERDA.pdf.

Thomspon, J. (2008). Uber Recycling. *High Country News, 4(11),* 8-9. September 9, 2008

USFN (2008). United States Federal News Service, Washington D.C. accessed 8-27-2008 through PROQUEST.

Valkenburg, C, Walton, C. W., Thompson, B. L., Gerber, M. A., Jones, S. & Stevens. D. J. (2008). *Municipal Solid Waste (MSW) to Liquid Fuels Synthesis, Volume 1: Availability of Feedstock and Technology.* PNNL 18144, Pacific Northwest National Laboratory, Richland, WA.

WAC 173-306-200 Washington Administrative Code 173-306-200 Generator Management Plans. *http://apps.leg.wa.gov/WAC/default. aspx?cite=173-306&full=true#173-306-200.*

Zhu, Y., Gerber, M. A., Jones, S. B. & Stevens, D. J. (2009). *Analysis of the Effects of Compositional and Configurational Assumptions on Product Costs for the Thermochemical Conversion of Lignocellulosic Biomass to Mixed Alcohols – FY 2007 Progress Report.* PNNL- 17949 Revision 1, Pacific Northwest National Laboratory, Richland, WA.

APPENDIX A. HEAT AND MATERIAL BALANCE FOR THE BIOMASS REFERENCE CASE

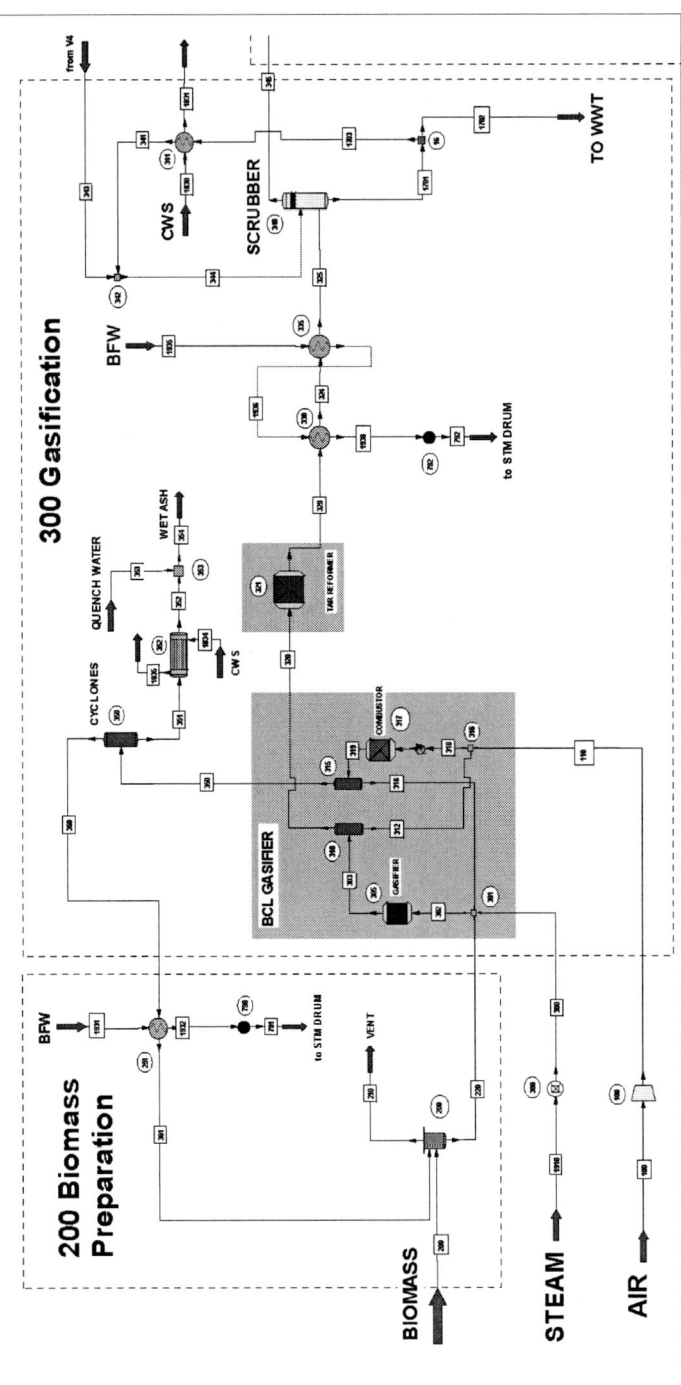

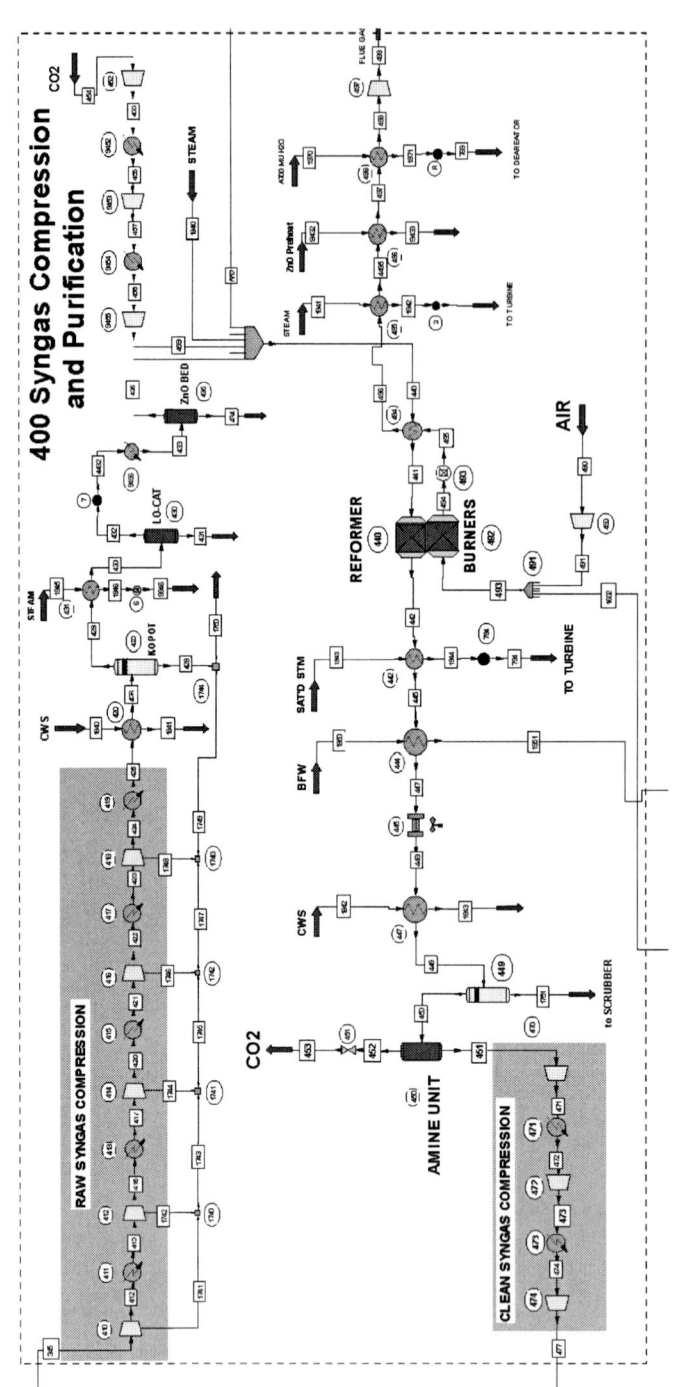

Figure 1. (Continued.).

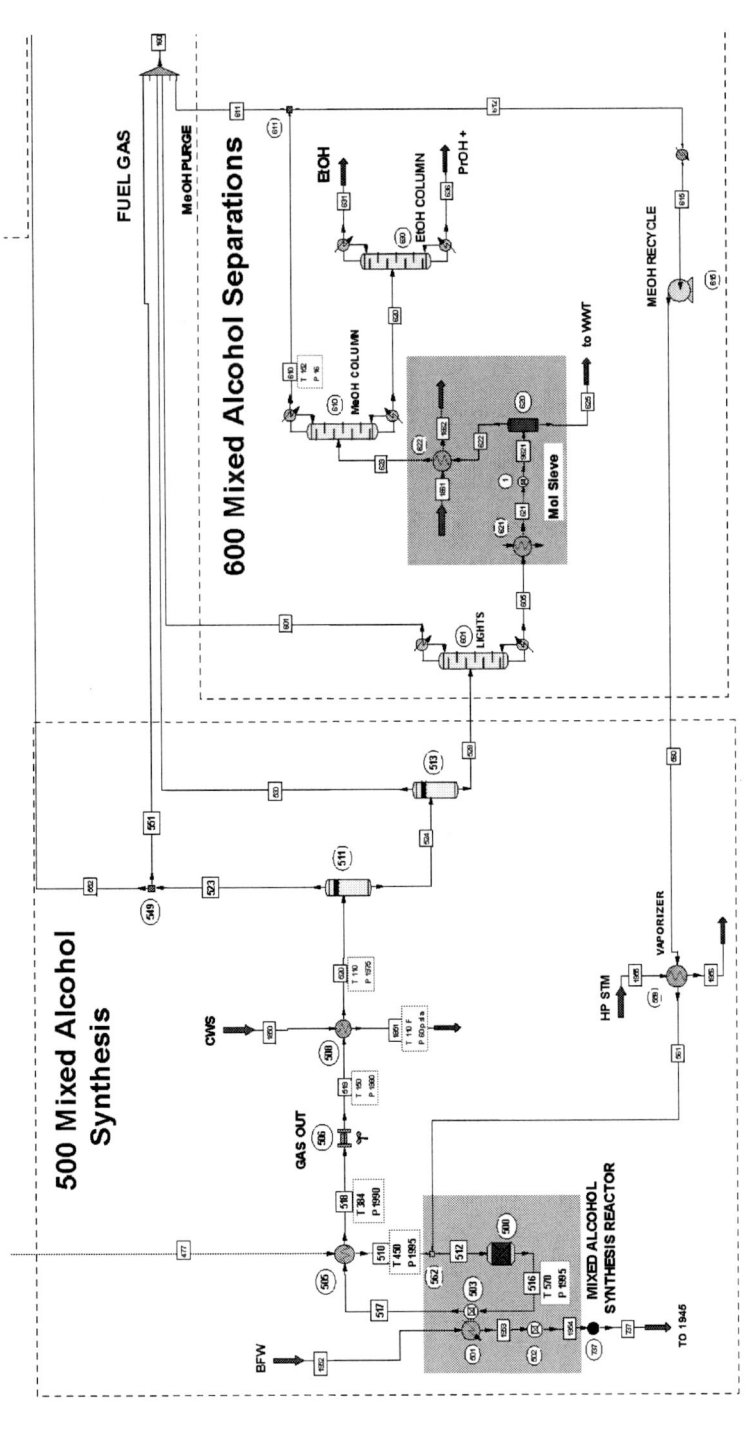

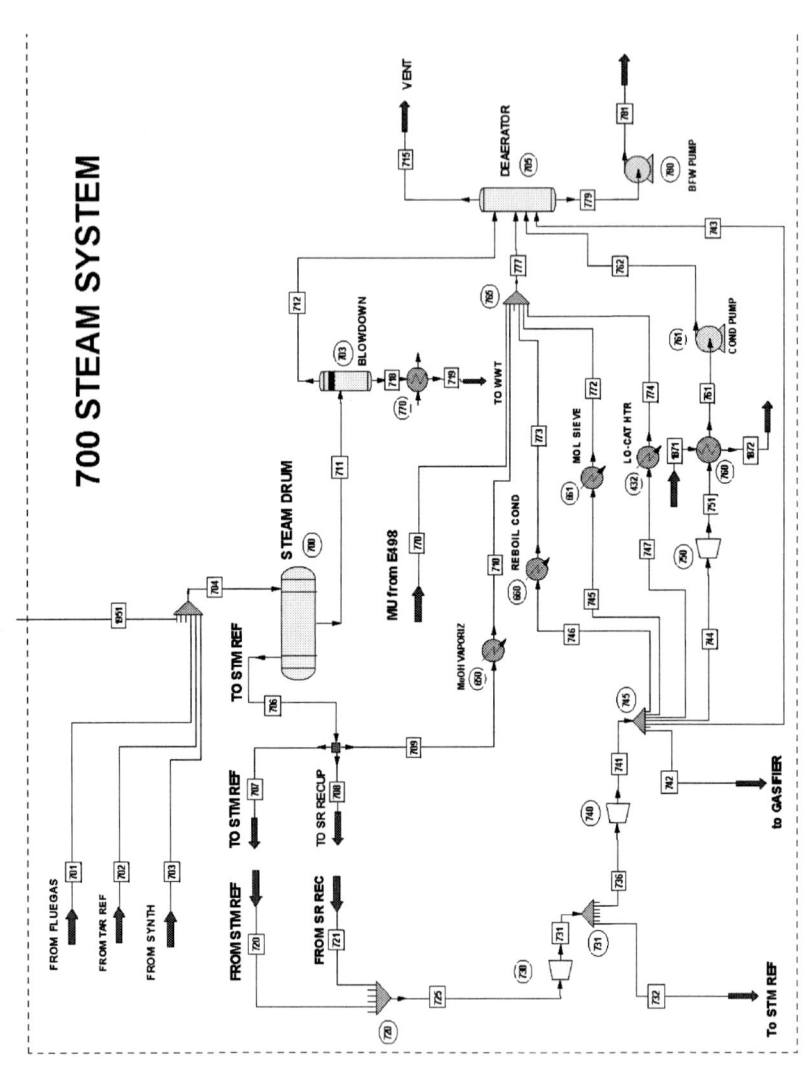

Figure A-1. Process Flow Diagram for the Biomass Reference Case.

Table A-1. Stream Results for the Biomass Reference Case

Stream No.	5	100	110	200	210	220	300	302	303	312	316	318	319	320	324
Stream Name		AIR	BIOMASS VENT									Syngas			
Temp F	1611.079	90	184.7633	60	225.2196	220	260	1737.218	1598	1598	1823	1523.3	1823	1598	464.2304
Pres psia	22	14.696	22	25	22	22	25	22	23	23	22	22	22	23	19
Enth MM Btu/h	-8518.2	-47.987	-38.01	-1692.5	-1470.1	-559.18	-416.82	-9151.3	-9341.2	-8670.1	-8175.3	-8708.1	-8509.2	-671.07	-756.76
Vapor mass fraction	1	1	1	0	1	0	1	0.33736	1	1	0	1	1	1	1
Total lb/h	5418386	430413	430413	367437.4	628558	204132	73119.59	5228696	5228662	4987973	4951445	5418386	5418387	240689.4	240689.1
Flowrates in lb/h															
Oxygen	102708.3	97654.13	97654.13	0	11292.91	0	0	0	5054.145	5054.145	0	102708.3	11292.91	0	0
Nitrogen	318713.3	318713.3	318713.3	0	318689	0	0	0	0	0	0	318713.3	318689	0	204.734
Argon	5435.613	5435.613	5435.613	0	5435.613	0	0	0	0	0	0	5435.613	5435.613	0	0
Carbon	27500.24	0	0	0	0	0	0	0	27500.24	27500.24	0	27500.24	0.0012	0	0
Hydrogen	2282.96	0	0	0	0.0009	0	0	0	5461.737	2282.96	0	2282.96	0.0009	3178.778	6370.468
Carbon Monoxide	0	0	0	0	0	0	0	0	78075.38	0	0	0	0	78075.38	98480.85
Carbon Dioxide	212.2069	212.2069	212.2069	0	100981	0	0	0	37202.6	0	0	212.2069	100981	37202.59	37202.6

Table A-1. (Continued)

Stream No.	5	100	110	200	210	220	300	302	303	312	316	318	319	320	324
Methane	0	0	0	0	0	0	0	0	16283.21	0	0	0	0	16283.21	13026.57
Acetylene	0	0	0	0	0	0	0	0	701.6728	0	0	0	0	701.6727	350.8364
Ethylene	0	0	0	0	0	0	0	0	8060.712	0	0	0	0	8060.713	4030.356
Ethane	0	0	0	0	0	0	0	0	579.8716	0	0	0	0	579.8715	57.9872
Propane	0	0	0	0	0	0	0	0	0	0	0	0	0	0	0
Water	8397.746	8397.746	8397.746	183718.8	192107.1	20413.18	73119.59	93532.76	93532.76	0	0	8397.746	28801.52	93532.77	80408.71
Sulphur	0	0	0	0	0	0	0	0	0	0	0	0	0	0	0
Carbonyl Sulfide	0	0	0	0	0	0	0	0	0	0	0	0	0	0	0
Hydrogen Sulfide	0	0	0	0	0	0	0	0	161.5833	0	0	0	0	161.5833	161.5833
Ammonia	0	0	0	0	0	0	0	0	355.6206	0	0	0	0	355.6205	106.6862
Hydrogen Chloride	0	0	0	0	0	0	0	0	0	0	0	0	0	0	0
Silicon Dioxide	4951445	0	0	0	0	0	0	4951445	4951445	4951445	4951445	4951445	4951445	0	0
Calcium Oxide	1690.211	0	0	0	0	0	0	0	1690.211	1690.211	0	1690.211	1690.222	0	0
Benzene	0	0	0	0	0	0	0	0	639.3002	0	0	0	0	639.3002	191.7901
Naphth-alene	0	0	0	0	0	0	0	0	1917.901	0	0	0	0	1917.901	95.8951
Hybrid Pop NREL	0.0041	0	0	183718.6	0	183718.6	0	183718.6	0.0041	0.0041	0	0.0041	0	0	0

Table A-1. (Continued)

Stream No.	5	100	110	200	210	220	300	302	303	312	316	318	319	320	324
Sulfur Dioxide	0	0	0	0	0	0	0	0	0	0	0	0	0	0	0
Hydrogen Cyanide	0	0	0	0	0	0	0	0	0	0	0	0	0	0	0
Nitric Oxide	0	0	0	0	52.0559	0	0	0	0	0	0	0	52.0559	0	0
Methanol	0	0	0	0	0	0	0	0	0	0	0	0	0	0	0
Ethanol	0	0	0	0	0	0	0	0	0	0	0	0	0	0	0
Isopropanol	0	0	0	0	0	0	0	0	0	0	0	0	0	0	0
N-Propanol	0	0	0	0	0	0	0	0	0	0	0	0	0	0	0
Isobutanol	0	0	0	0	0	0	0	0	0	0	0	0	0	0	0
N-Butanol	0	0	0	0	0	0	0	0	0	0	0	0	0	0	0
1-Pentanol	0	0	0	0	0	0	0	0	0	0	0	0	0	0	0

Stream No.	325	329	341	343	344	345	350	351	352	353	354	360	361	412	413
Stream Name			from V449					ASH		QUENCH WWET ASH					
Temp F	250.8848	1383	110	110	109.9997	118.317	1823	1823	300	60	137.5378	1823	1750	253.0462	140
Pres psia	18	20	15	415	15	15	22	22	22	14.7	14.7	22	22	28	28
Enth MM Btu/h	-778.95	-651	-66654	-253.51	-66907	-435.29	-333.87	-7.5742	-8.1505	-5.1321	-13.283	-326.3	-336.8	-425.53	-434.68
Vapor mass fraction	1	1	0.002532	0	0.002522	1	1	0	0	0	0	1	1	1	0.99495
Total lb/h	240689.1	240689.1	9826158	37319.66	9863478	178568.7	466942.3	1690.224	1690.224	750.0001	2440.224	465252.1	465252.1	178568.7	178568.7
Flowrates in lb/h															
Oxygen	0	0	0	0	0	0	11292.91	0	0	0	0	11292.91	11292.91	0	0

Table A-1. (Continued)

Stream No.	325	329	341	343	344	345	350	351	352	353	354	360	361	412	413
Nitrogen	204.734	204.734	0.0907	0	0.0907	204.733	318689	0	0	0	0	318689	318689	204.733	204.733
Argon	0	0	0	0	0	0	5435.613	0	0	0	0	5435.613	5435.613	0	0
Carbon	0	0	0	0	0	0	0.0012	0.0012	0.0012	0	0.0012	0	0	0	0
Hydrogen	6370.468	6370.468	4.0767	0	4.0767	6370.427	0.0009	0	0	0	0	0.0009	0.0009	6370.427	6370.427
Carbon Monoxide	98480.85	98480.85	64.4907	0	64.4907	98480.19	0	0	0	0	0	0	0	98480.19	98480.19
Carbon Dioxide	37202.6	37202.6	17673.67	0	17673.67	36987.35	100981	0	0	0	0	100981	100981	36987.35	36987.35
Methane	13026.57	13026.57	13.5323	0	13.5323	13026.43	0	0	0	0	0	0	0	13026.43	13026.43
Acetylene	350.8364	350.8364	0.3639	0	0.3639	350.8326	0	0	0	0	0	0	0	350.8326	350.8326
Ethylene	4030.356	4030.356	13.0743	0	13.0743	4030.22	0	0	0	0	0	0	0	4030.222	4030.222
Ethane	57.9872	57.9872	0.0641	0	0.0641	57.9865	0	0	0	0	0	0	0	57.9865	57.9865
Propane	0	0	0	0	0	0	0	0	0	0	0	0	0	0	0
Water	80408.71	80408.71	9801370	37319.66	9838689	18585	28801.52	0	0	750.0001	750.0001	28801.52	28801.52	18585	18585
Sulphur	0	0	0	0	0	0	0	0	0	0	0	0	0	0	0
Carbonyl Sulfide	0	0	0	0	0	0	0	0	0	0	0	0	0	0	0
Hydrogen Sulfide	161.5833	161.5833	108.8491	0	108.8491	160.6335	0	0	0	0	0	0	0	160.6335	160.6335
Ammonia	106.6862	106.6862	6909.775	0	6909.775	27.2479	0	0	0	0	0	0	0	27.2479	27.2479
Hydrogen Chloride	0	0	0	0	0	0	0	0	0	0	0	0	0	0	0

Table A-1. (Continued)

Stream No.	325	329	341	343	344	345	350	351	352	353	354	360	361	412	413
Silicon Dioxide	0	0	0	0	0	0	0	0	0	0	0	0	0	0	0
Calcium Oxide	0	0	0	0	0	0	1690.222	1690.222	1690.222	0	1690.222	0	0	0	0
Benzene	191.7901	191.7901	0.2888	0	0.2888	191.787	0	0	0	0	0	0	0	191.787	191.787
Naphthalene	95.8951	95.8951	0.0851	0	0.0851	95.8941	0	0	0	0	0	0	0	95.8941	95.8941
Hybrid Pop NREL	0	0	0	0	0	0	0	0	0	0	0	0	0	0	0
Sulfur Dioxide	0	0	0	0	0	0	0	0	0	0	0	0	0	0	0
Hydrogen Cyanide	0	0	0	0	0	0	0	0	0	0	0	0	0	0	0
Nitric Oxide	0	0	0	0	0	0	52.0559	0	0	0	0	52.0559	52.0559	0	0
Methanol	0	0	0	0	0	0	0	0	0	0	0	0	0	0	0
Ethanol	0	0	0	0	0	0	0	0	0	0	0	0	0	0	0
Isopropanol	0	0	0	0	0	0	0	0	0	0	0	0	0	0	0
N-Propanol	0	0	0	0	0	0	0	0	0	0	0	0	0	0	0
Isobutanol	0	0	0	0	0	0	0	0	0	0	0	0	0	0	0
N-Butanol	0	0	0	0	0	0	0	0	0	0	0	0	0	0	0
1-Pentanol	0	0	0	0	0	0	0	0	0	0	0	0	0	0	0
Stream No.	416	417	420	421	422	423	424	425	426	428	429	430	431	432	433
Stream Name															

Table A-1. (Continued)

Stream No.	416	417	420	421	422	423	424	425	426	428	429	430	431	432	433
Temp F	274.5935	140	195.4107	140	209.0953	140	231.2065	140	110	110	110	120	120	120	707
Pres psia	54	54	109.5	109.5	220	220	465	465	465	465	465	465	445	445	445
Enth MMBtu/h	-423.96	-443.82	-416.31	-432.43	-381.76	-394.84	-364.62	-377.55	-380.31	-48.377	-331.94	-331.28	0	-331.26	-291.04
Vapor mass fraction	1	0.94457	1	0.9317	1	0.95183	1	0.9609	0.9575	0	1	1	0	1	1
Total lb/h	178568.7	178568.7	176201.6	176201.6	170321.2	170321.2	167569.5	167569.5	167569.5	7121.836	160447.6	160447.6	0	160447.6	160447.6
Flowrates in lb/h															
Oxygen	0	0	0	0	0	0	0	0	0	0	0	0	0	0	0
Nitrogen	204.733	204.733	204.733	204.733	204.733	204.733	204.733	204.733	204.733	0	204.733	204.733	0	204.733	204.733
Argon	0	0	0	0	0	0	0	0	0	0	0	0	0	0	0
Carbon	0	0	0	0	0	0	0	0	0	0	0	0	0	0	0
Hydrogen	6370.427	6370.427	6370.427	6370.427	6370.427	6370.427	6370.426	6370.426	6370.426	0	6370.426	6370.426	0	6370.426	6370.426
Carbon Monoxide	98480.19	98480.19	98483.19	98480.19	98480.19	98480.19	98480.19	98480.19	98480.19	0	98480.19	98480.19	0	98480.19	98480.19
Carbon Dioxide	36987.35	36987.35	36987.35	36987.35	36987.35	36987.35	36987.35	36987.35	36987.35	0	36987.35	36987.35	0	36987.35	36987.35
Methane	13026.43	13026.43	13025.43	13026.43	13026.43	13026.43	13026.43	13026.43	13026.43	0	13026.43	13026.43	0	13026.43	13026.43
Acetylene	350.8326	350.8326	350.8326	350.8326	350.8326	350.8326	350.8326	350.8326	350.8326	0	350.8326	350.8326	0	350.8326	350.832
Ethylene	4030.222	4030.222	4030.222	4030.222	4030.222	4030.222	4030.222	4030.222	4030.222	0	4030.222	4030.222	0	4030.222	4030.222
Ethane	57.9865	57.9865	57.9865	57.9865	57.9865	57.9865	57.9865	57.9865	57.9865	0	57.9865	57.9865	0	57.9865	57.9865

Table A-1. (Continued)

Stream No.	416	417	420	421	422	423	424	425	426	428	429	430	431	432	433
Propane	0	0	0	0	0	0	0	0	0	0	0	0	0	0	0
Water	18585	18585	16217.9	16217.9	10337.51	10337.51	7585.755	7585.755	7585.755	7121.836	463.9131	463.9131	0	463.9131	463.9131
Sulphur	0	0	0	0	0	0	0	0	0	0	0	0	0	0	0
Carbonyl Sulfide	0	0	0	0	0	0	0	0	0	0	0	0	0	0	0
Hydrogen Sulfide	160.6335	160.6335	160.6335	160.6335	160.6335	160.6335	160.6335	160.6335	160.6335	0	160.6335	160.6335	0	160.6335	160.6335
Ammonia	27.2479	27.2479	27.2479	27.2479	27.2479	27.2479	27.2479	27.2479	27.2479	0	27.2479	27.2479	0	27.2479	27.2479
Hydrogen Chloride	0	0	0	0	0	0	0	0	0	0	0	0	0	0	0
Silicon Dioxide	0	0	0	0	0	0	0	0	0	0	0	0	0	0	0
Calcium Oxide	0	0	0	0	0	0	0	0	0	0	0	0	0	0	0
Benzene	191.787	191.787	191.787	191.787	191.787	191.787	191.787	191.787	191.787	0	191.787	191.787	0	191.787	191.787
Naphthalene	95.8941	95.8941	95.8941	95.8941	95.8941	95.8941	95.8941	95.8941	95.8941	0	95.8941	95.8941	0	95.8941	95.8941
Hybrid Pop NREL	0	0	0	0	0	0	0	0	0	0	0	0	0	0	0
Sulfur Dioxide	0	0	0	0	0	0	0	0	0	0	0	0	0	0	0
Hydrogen Cyanide	0	0	0	0	0	0	0	0	0	0	0	0	0	0	0
Nitric Oxide	0	0	0	0	0	0	0	0	0	0	0	0	0	0	0
Methanol	0	0	0	0	0	0	0	0	0	0	0	0	0	0	0

Table A-1. (Continued)

Stream No.	416	417	420	421	422	423	424	425	426	428	429	430	431	432	433
Ethanol	0	0	0	0	0	0	0	0	0	0	0	0	0	0	0
Isopropanol	0	0	0	0	0	0	0	0	0	0	0	0	0	0	0
N-Propanol	0	0	0	0	0	0	0	0	0	0	0	0	0	0	0
Isobutanol	0	0	0	0	0	0	0	0	0	0	0	0	0	0	0
N-Butanol	0	0	0	0	0	0	0	0	0	0	0	0	0	0	0
1-Pentanol	0	0	0	0	0	0	0	0	0	0	0	0	0	0	0
Stream No.	434	435	440	441	442	445	447	448	449	450	451	452	453	454	455
Stream Name												CO_2		CO_2	
Temp F	707.0002	707	459.8756	930.0002	1652	1170.387	300.0607	150	110	110	120	120	57.4546	60	316.153
Pres psia	682	682	450	445	430	430	430	430	427.5	427.5	422.5	422.5	22	22	100
Enth MM Btu/h	-0.01686	-290.95	-1947.7	-1828.7	-1527	-1663	-1927.1	-2042.1	-2056.9	-1236.2	-432.73	-803.75	-803.75	-609.7	-601
Vapor mass fraction	1	1	1	1	1	1	0.92981	0.80342	0.79916	1	1	0.995	0.99766	1	1
Total lb/h	160.6335	160287	601582.1	601582	601582	601582	601582	601582	601582	480759	272761.6	207997.1	207997.1	158436	158436
Flowrates in lb/h															
Oxygen	0	0	0	0	0	0	0	0	0	0	0	0	0	0	0
Nitrogen	0	204.733	667.2862	667.2863	680.811	680.811	680.811	680.811	680.811	680.811	680.811	0	0	0	0

Table A-1. (Continued)

Stream No.	434	435	440	441	442	445	447	448	449	450	451	452	453	454	455
Argon	0	0	0	0	0	0	0	0	0	0	0	0	0	0	0
Carbon	0	0	0	0	0	0	0	0	0	0	0	0	0	0	0
Hydrogen	0	6370.426	13637.61	13637.61	19134.34	19134.34	19134.34	19134.34	19134.34	19134.34	19134.34	0	0	0	0
Carbon Monoxide	0	98480.19	191381.6	191381.6	235921.5	235921.5	235921.5	235921.5	235921.5	235921.4	235921.4	0	0	0	0
Carbon Dioxide	0	36987.35	226418.7	226418.7	208683.4	208683.4	208683.4	208683.4	208683.4	208683.4	2086.822	206596.6	206596.6	158436	158436
Methane	0	13026.43	27171.88	27171.89	14917.88	14917.88	14917.88	14917.88	14917.88	14917.88	14917.88	0	0	0	0
Acetylene	0	350.8326	350.8344	350.8345	0.0031	0.0031	0.0031	0.0031	0.0031	0.0031	0.0031	0	0	0	0
Ethylene	0	4030.222	4030.67	4030.67	0.6849	0.6849	0.6849	0.6849	0.6849	0.6849	0.6849	0	0	0	0
Ethane	0	57.9865	469.5271	469.5272	1.5415	1.5415	1.5415	1.5415	1.5415	1.5415	1.5415	0	0	0	0
Propane	0	0	0	0	0	0	0	0	0	0	0	0	0	0	0
Water	0	463.9131	135730.5	135730.4	122223.7	122223.7	122223.7	122223.7	122223.7	1400.485	0	1400.485	1400.48	0	0
Sulphur	0	0	0	0	0	0	0	0	0	0	0	0	0	0	0
Carbonyl Sulfide	0	0	0	0	0	0	0	0	0	0	0	0	0	0	0
Hydrogen Sulfide	160.6335	0	0	0	0	0	0	0	0	0	0	0	0	0	0
Ammonia	0	27.2479	34.2377	34.2377	17.6516	17.6516	17.6516	17.6516	17.6516	17.6516	17.6516	0	0	0	0
Hydrogen Chloride	0	0	0	0	0	0	0	0	0	0	0	0	0	0	0
Silicon Dioxide	0	0	0	0	0	0	0	0	0	0	0	0	0	0	0
Calcium Oxide	0	0	0	0	0	0	0	0	0	0	0	0	0	0	0

Table A-1. (Continued)

Stream No.	434	435	440	441	442	445	447	448	449	450	451	452	453	454	455
Benzene	0	191.787	191.7869	191.787	0	0	0	0	0	0	0	0	0	0	0
Naphthalene	0	95.8941	95.8941	95.8941	0	0	0	0	0	0	0	0	0	0	0
Hybrid Pop NREL	0	0	0	0	0	0	0	0	0	0	0	0	0	0	0
Sulfur Dioxide	0	0	0	0	0	0	0	0	0	0	0	0	0	0	0
Hydrogen Cyanide	0	0	0.0618	0.0618	0.2912	0.2912	0.2912	0.2912	0.2912	0.2912	0.2912	0	0	0	0
Nitric Oxide	0	0	0	0	0	0	0	0	0	0	0	0	0	0	0
Methanol	0	0	492.663	492.6631	0.1196	0.1196	0.1196	0.1196	0.1196	0.1196	0.1196	0	0	0	0
Ethanol	0	0	792.6373	792.6376	0.0001	0.0001	0.0001	0.0001	0.0001	0.0001	0.0001	0	0	0	0
Isopropanol	0	0	0	0	0	0	0	0	0	0	0	0	0	0	0
N-Propanol	0	0	87.9436	87.9436	0	0	0	0	0	0	0	0	0	0	0
Isobutanol	0	0	0	0	0	0	0	0	0	0	0	0	0	0	0
N-Butanol	0	0	19.1715	19.1715	0	0	0	0	0	0	0	0	0	0	0
1-Pentanol	0	0	8.9839	8.9839	0	0	0	0	0	0	0	0	0	0	0
Stream No.	456	457	458	459	471	472	473	474	477	490	491	493	494	495	496
Stream Name										AIR			To LOCAT		
Temp F	150	317.6949	150	256.3384	235.4644	150	277.7088	150	270.7586	60	124.5558	111.9801	1800	1800	
Pres psia	100	250	250	450	700	700	1200	1200	2000	14.696	20	16	16	16	
Enth MM Btu/h	-607	-601.26	-607.55	-604.12	-416.99	-428.73	-411	-428.8	-411.63	-1.7701	4.8091	-166.6	-470.27	-470.27	

Table A-1. (Continued)

Stream No.	456	457	458	459	471	472	473	474	477	490	491	493	494	495	496
Vapor mass fraction	1	1	1	1	1	1	1	1	1	1	1	1	1	1	1
Total lb/h	158436	158436	158436	158436	272761.6	272761.6	272761.6	272761.6	272761.6	421657.2	421657.2	498803.3	498803.5	498803.5	498803.5
Flowrates in lb/h															
Oxygen	0	0	0	0	0	0	0	0	0	98210	98210	98210	12717.52	12717.52	12717.52
Nitrogen	0	0	0	0	680.811	680.811	680.811	680.811	680.811	323447.3	323447.3	323666	323651	323651	323651
Argon	0	0	0	0	0	0	0	0	0	0	0	0	0	0	0
Carbon	0	0	0	0	0	0	0	0	0	0	0	0	0	0	0
Hydrogen	0	0	0	0	19134.34	19134.34	19134.34	19134.34	19134.34	0	0	3434.957	0.0013	0.0013	0.0013
Carbon Monoxide	0	0	0	0	235921.4	235921.4	235921.4	235921.4	235921.4	0	0	44032.44	0.028	0.028	0.028
Carbon Dioxide	158436	158436	158436	158436	2086.822	2086.822	2086.822	2086.822	2086.822	0	0	19273.77	112381	112381	112381
Methane	0	0	0	0	14917.88	14917.88	14917.88	14917.88	14917.88	0	0	6844.954	0.0003	0.0003	0.0003
Acetylene	0	0	0	0	0.0031	0.0031	0.0031	0.0031	0.0031	0	0	0.0012	0.0005	0.0005	0.0005
Ethylene	0	0	0	0	0.6849	0.6849	0.6849	0.6849	0.6849	0	0	0.2364	0.0005	0.0005	0.0005
Ethane	0	0	0	0	1.5415	1.5415	1.5415	1.5415	1.5415	0	0	223.0892	0.0006	0.0006	0.0006
Propane	0	0	0	0	0	0	0	0	0	0	0	0	0	0	0
Water	0	0	0	0	0	0	0	0	0	0	0	126.0698	50003	50003	50003
Sulphur	0	0	0	0	0	0	0	0	0	0	0	0	0	0	0
Carbonyl Sulfide	0	0	0	0	0	0	0	0	0	0	0	0	0	0	0
Hydrogen Sulfide	0	0	0	0	0	0	0	0	0	0	0	0	0	0	0

Table A-1. (Continued)

Stream No.	456	457	458	459	471	472	473	474	477	490	491	493	494	495	496
Ammonia	0	0	0	0	17.6516	17.6516	17.6516	17.6516	17.6516	0	0	10.6638	0.0003	0.0003	0.0003
Hydrogen Chloride	0	0	0	0	0	0	0	0	0	0	0	0	0	0	0
Silicon Dioxide	0	0	0	0	0	0	0	0	0	0	0	0	0	0	0
Calcium Oxide	0	0	0	0	0	0	0	0	0	0	0	0	0	0	0
Benzene	0	0	0	0	0	0	0	0	0	0	0	0	0	0	0
Naphthalene	0	0	0	0	0	0	0	0	0	0	0	0	0	0	0
Hybrid Pop NREL	0	0	0	0	0	0	0	0	0	0	0	0	0	0	0
Sulfur Dioxide	0	0	0	0	0	0	0	0	0	0	0	0	0	0	0
Hydrogen Cyanide	0	0	0	0	0.2912	0.2912	0.2912	0.2912	0.2912	0	0	0.2293	0.0005	0.0005	0.0005
Nitric Oxide	0	0	0	0	0	0	0	0	0	0	0	0	51.0227	51.0227	51.0227
Methanol	0	0	0	0	0.1196	0.1196	0.1196	0.1196	0.1196	0	0	2293.303	0.0006	0.0006	0.0006
Ethanol	0	0	0	0	0.0001	0.0001	0.0001	0.0001	0.0001	0	0	609.9102	0.0009	0.0009	0.0009
Isopropanol	0	0	0	0	0	0	0	0	0	0	0	0	0	0	0
N-Propanol	0	0	0	0	0	0	0	0	0	0	0	59.8768	0.0011	0.0011	0.0011
Isobutanol	0	0	0	0	0	0	0	0	0	0	0	0	0	0	0
N-Butanol	0	0	0	0	0	0	0	0	0	0	0	12.3067	0.0014	0.0014	0.0014
1-Pentanol	0	0	0	0	0	0	0	0	0	0	0	5.6418	0.0016	0.0016	0.0016
Stream No.	497	498	499	510	512	516	517	518	519	520	523	524	528	530	551
Stream Name	FLUE GAS					GAS OUT				TO WWT					FLUE GAS
Temp F	469.7888	280	297.8397	450	451.0664	570	570	383.6532	150	110	110.2585	110.2585	101.7292		469.7888

Table A-1. (Continued)

Stream No.	497	498	499	510	512	516	517	518	519	520	523	524	528	530	551
Pres psia	14	14	15	1995	1995	1995	1995	1990	1980	1975	1975	1975	35	35	1975
Enth MMBtu/h	-668.94	-694.37	-692.01	-386.4	-418.49	-580.02	-580.02	-605.26	-657.23	-664.53	-454.55	-209.97	-193.71	-16.261	-145.46
Vapor mass fraction	1	1	1	1	1	1	1	1	0.78186	0.76503	1	0	0	1	1
Total lb/h	498803.5	498803.5	498803.5	272761.6	284142.6	284144.6	284144.6	284144.6	284144.6	284144.6	217274.2	66870.41	62228.29	4642.12	69527.73
Flowrates in lb/h															
Oxygen	12717.52	12717.52	12717.52	0	0	0	0	0	0	0	0	0	0	0	0
Nitrogen	323651	323651	323651	680.811	681.2112	681.211	681.211	681.211	681.211	681.211	680.2254	0.9856	0.0024	0.9832	217.6721
Argon	0	0	0	0	0	0	0	0	0	0	0	0	0	0	0
Carbon	0	0	0	0	0	0	0	0	0	0	0	0	0	0	0
Hydrogen	0.0013	0.0013	0.0013	19134.34	19130.62	10702.14	10702.14	10702.14	10702.14	10702.14	10687.04	15.1045	0.0318	15.0728	3419.852
Carbon Monoxide	0.028	0.028	0.028	235921.5	235887.4	136934	136934	136934	136934	136934	136619.8	314.1132	1.2259	312.8874	43718.33
Carbon Dioxide	112381	112381	112381	2086.822	2086.908	50269.07	50269.07	50269.07	50269.07	50269.07	45581.32	4687.747	986.0571	3701.69	14586.02
Methane	0.0003	0.0003	0.0003	14917.89	14910.61	20990.42	20990.42	20990.42	20990.42	20990.42	20802.15	188.2663	3.3535	184.9128	6656.688
Acetylene	0.0005	0.0005	0.0005	0.0031	0.0031	0.0031	0.0031	0.0031	0.0031	0.0031	0.0028	0.0002	0	0.0002	0.0009
Ethylene	0.0005	0.0005	0.0005	0.6849	0.6844	0.6844	0.6844	0.6844	0.6844	0.6844	0.6589	0.0256	0.0023	0.0233	0.2108
Ethane	0.0006	0.0006	0.0006	1.5415	1.5403	634.6297	634.6297	634.6297	634.6297	634.6297	605.2066	29.4231	3.3546	26.0685	193.6661
Propane	0	0	0	0	0	0	0	0	0	0	0	0	0	0	0
Water	50003	50003	50003	0	0	7281.481	7281.481	7281.481	7281.481	7281.481	226.5492	7054.934	7001.711	53.2222	72.4957

Table A-1. (Continued)

Stream No.	497	498	499	510	512	516	517	518	519	520	523	524	528	530	551
Sulphur	0	0	0	0	0	0	0	0	0	0	0	0	0	0	0
Carbonyl Sulfide	0	0	0	0	0	0	0	0	0	0	0	0	0	0	0
Hydrogen Sulfide	0	0	0	0	0	0	0	0	0	0	0	0	0	0	0
Ammonia	0.0003	0.0003	0.0003	17.6516	17.6536	17.6537	17.6537	17.6537	17.6537	17.6537	10.2791	7.3745	5.0573	2.3172	3.2893
Hydrogen Chloride	0	0	0	0	0	0	0	0	0	0	0	0	0	0	0
Silicon Dioxide	0	0	0	0	0	0	0	0	0	0	0	0	0	0	0
Calcium Oxide	0	0	0	0	0	0	0	0	0	0	0	0	0	0	0
Benzene	0	0	0	0	0	0	0	0	0	0	0	0	0	0	0
Naphthalene	0	0	0	0	0	0	0	0	0	0	0	0	0	0	0
Hybrid Pop NREL	0	0	0	0	0	0	0	0	0	0	0	0	0	0	0
Sulfur Dioxide	0	0	0	0	0	0	0	0	0	0	0	0	0	0	0
Hydrogen Cyanide	0.0005	0.0005	0.0005	0.2912	0.2912	0.2912	0.2912	0.2912	0.2912	0.2912	0.0909	0.2002	0.1813	0.0189	0.0291
Nitric Oxide	51.0227	51.0227	51.0227	0	0	0	0	0	0	0	0	0	0	0	0
Methanol	0.0006	0.0006	0.0006	0.1196	11398.28	14312.07	14312.07	14312.07	14312.07	14312.07	724.5045	13587.57	13465.86	121.7069	231.8414
Ethanol	0.0009	0.0009	0.0009	0.0001	27.1554	31646.65	31646.65	31646.65	31646.65	31646.65	1165.643	30481.02	30280.96	200.0588	373.0058
Isopropanol	0	0	0	0	0	0	0	0	0	0	0	0	0	0	0
N-Propanol	0.0011	0.0011	0.0011	0	0.2859	6557.749	6557.749	6557.749	6557.749	6557.749	129.3288	6428.421	6409.961	18.4592	41.3852

Table A-1. (Continued)

Stream No.	497	498	499	510	512	516	517	518	519	520	523	524	528	530	551
Isobutanol	0	0	0	0	0	0	0	0	0	0	0	0	0	0	0
N-Butanol	0.0014	0.0014	0.0014	0	0	2747.122	2747.122	2747.122	2747.122	2747.122	28.1934	2718.929	2715.644	3.2848	9.0219
1-Pentanol	0.0016	0.0016	0.0016	0	0	1369.51	1369.51	1369.51	1369.51	1369.51	13.2117	1356.298	1354.884	1.414	4.2277

Stream No.	552	560	561	601	605	610	611	612	615	620	621	22	623	625	631
Stream Name		MeOH			MeOH		PURMeOH RECYCLE			Mixed OH		to WWT			EtOH
Temp F	110.2585	167.2547	480	136.2132	194.8936	152.0138	152.0138	152.0138	152.0138	197.9772	193.9325	193.9326	187.4027	193.9326	176.7088
Pres psia	1975	2000	1995	23	26.7	16	16	16	16	22	26.7	23	23	23	16
Enth MMBtu/h	-309.1	-35.947	-32.154	-5.6871	-183.46	-40.074	-4.0074	-36.067	-36.067	-96.767	-157.24	-132.03	-136.85	-46.917	-76.788
Vapor mass fraction	1	0	1	1	0	0	0	0	0	0	1	0.2243	2.85E-06	0	0
Total lb/h	147746.4	11425.6	11425.6	1706.739	60521.55	12695.11	1269.511	11425.6	11425.6	40825.07	60521.55	53520.18	53520.18	7001.335	30393.73
Flowrates in lb/h															
Oxygen	0	0	0	0	0	0	0	0	0	0	0	0	0	0	0
Nitrogen	462.5533	0	0	0.0024	0	0	0	0	0	0	0	0	0	0	0
Argon	0	0	0	0	0	0	0	0	0	0	0	0	0	0	0
Carbon	0	0	0	0	0	0	0	0	0	0	0	0	0	0	0
Hydrogen	7267.187	0	0	0.0318	0	0	0	0	0	0	0	0	0	0	0
Carbon Monoxide	92901.45	0	0	1.2259	0	0	0	0	0	0	0	0	0	0	0

Table A-1. (Continued)

Stream No.	552	560	561	601	605	610	611	612	615	620	621	22	623	625	631
Carbon Dioxide	30995.3	0	0	986.0571	0	0	0	0	0	0	0	0	0	0	0
Methane	14145.46	0	0	3.3535	0	0	0	0	0	0	0	0	0	0	0
Acetylene	0.0019	0	0	0	0	0	0	0	0	0	0	0	0	0	0
Ethylene	0.448	0	0	0.0023	0	0	0	0	0	0	0	0	0	0	0
Ethane	411.54 05	0	0	3.3546	0	0	0	0	0	0	0	0	0	0	0
Propane	0	0	0	0	0	0	0	0	0	0	0	0	0	0	0
Water	154.05 34	0	0	0.3518	7001.359	0	0	0	0	0	7001.3 59	0	0	7001.335	0
Sulphur	0	0	0	0	0	0	0	0	0	0	0	0	0	0	0
Carbonyl Sulfide	0	0	0	0	0	0	0	0	0	0	0	0	0	0	0
Hydrogen Sulfide	0	0	0	0	0	0	0	0	0	0	0	0	0	0	0
Ammonia	6.9898	0	0	5.0573	0	0	0	0	0	0	0	0	0	0	0
Hydrogen Chloride	0	0	0	0	0	0	0	0	0	0	0	0	0	0	0
Silicon Dioxide	0	0	0	0	0	0	0	0	0	0	0	0	0	0	0
Calcium Oxide	0	0	0	0	0	0	0	0	0	0	0	0	0	0	0
Benzene	0	0	0	0	0	0	0	0	0	0	0	0	0	0	0
Naphthalene	0	0	0	0	0	0	0	0	0	0	0	0	0	0	0
Hybrid Pop NREL	0	0	0	0	0	0	0	0	0	0	0	0	0	0	0
Sulfur Dioxide	0	0	0	0	0	0	0	0	0	0	0	0	0	0	0

Table A-1. (Continued)

Stream No.	552	560	561	601	605	610	611	612	615	620	621	22	623	625	631
Hydrogen Cyanide	0.0618	0	0	0.1813	0	0	0	0	0	0	0	0	0	0	0
Nitric Oxide	0	0	0	0	0	0	0	0	0	0	0	0	0	0	0
Methanol	492.6631	11398.16	11398.16	673.2924	12792.57	12664.62	1266.462	11398.16	11398.16	127.9357	12792.57	12792.56	12792.56	0	127.7901
Ethanol	792.6373	27.1553	27.1553	33.8283	30247.13	30.1726	3.0173	27.1553	27.1553	30216.97	30247.13	30247.14	30247.14	0	30201.85
Isopropanol	0	0	0	0	0	0	0	0	0	0	0	0	0	0	0
N-Propanol	87.9436	0.2859	0.2859	0.0007	6409.961	0.3176	0.0318	0.2859	0.2859	6409.638	6409.961	6409.955	6409.955	0	64.0946
Isobutanol	0	0	0	0	0	0	0	0	0	0	0	0	0	0	0
N-Butanol	19.1715	0	0	0	2715.644	0	0	0	0	2715.64	2715.644	2715.64	2715.64	0	0
1-Pentanol	8.9839	0	0	0	1354.884	0	0	0	0	1354.882	1354.884	1354.882	1354.882	0	0

Stream No.	636	701	702	703	706	707	708	710	711	711	712	712	715	718	719
Stream Name	PrOH +	FROM SYNTH	FROM FLU	FROM TAR									VENT	TO WWT	
Temp F	230.5994	526.5776	526.5776	526.5776	526.5776	526.5776	526.5776	526.5776	526.5776	526.5776	526.5776	526.5755	236.9942	526.5755	200
Pres psia	19	860	860	860	860	860	860	860	860	860	860	860	30	860	860
Enth MMBtu/h	-20.195	-60.648	-738.52	-932.53	-3256	-3219.3	-2468.7	-717.75	-32.91	-36.81	-36.695	0	0	-36.695	-38.749
Vapor mass fraction	0	0.99	0.99	0.98971	0.98991	1	1	1	1	0.001037	0	1	1	0	0
Total lb/h	10431.33	10675.7	130000	164146	573150	567369	435075	126494	5800	5800	5781.162	0	0	5781.162	5781.162
Flowrates in lb/h															

Table A-1. (Continued)

Stream No.	636	701	702	703	706	707	708	710	711	711	712	712	715	718	719
Oxygen	0	0	0	0	0	0	0	0	0	0	0	0	0	0	0
Nitrogen	0	0	0	0	0	0	0	0	0	0	0	0	0	0	0
Argon	0	0	0	0	0	0	0	0	0	0	0	0	0	0	0
Carbon	0	0	0	0	0	0	0	0	0	0	0	0	0	0	0
Hydrogen	0	0	0	0	0	0	0	0	0	0	0	0	0	0	0
Carbon Monoxide	0	0	0	0	0	0	0	0	0	0	0	0	0	0	0
Carbon Dioxide	0	0	0	0	0	0	0	0	0	0	0	0	0	0	0
Methane	0	0	0	0	0	0	0	0	0	0	0	0	0	0	0
Acetylene	0	0	0	0	0	0	0	0	0	0	0	0	0	0	0
Ethylene	0	0	0	0	0	0	0	0	0	0	0	0	0	0	0
Ethane	0	0	0	0	0	0	0	0	0	0	0	0	0	0	0
Propane	0	0	0	0	0	0	0	0	0	0	0	0	0	0	0
Water	0	10675.7	130000	164146	573150	567369	435075	126494	5800	5800	5781.162	0	0	5781.162	5781.162
Sulphur	0	0	0	0	0	0	0	0	0	0	0	0	0	0	0
Carbonyl Sulfide	0	0	0	0	0	0	0	0	0	0	0	0	0	0	0
Hydrogen Sulfide	0	0	0	0	0	0	0	0	0	0	0	0	0	0	0
Ammonia	0	0	0	0	0	0	0	0	0	0	0	0	0	0	0
Hydrogen Chloride	0	0	0	0	0	0	0	0	0	0	0	0	0	0	0
Silicon Dioxide	0	0	0	0	0	0	0	0	0	0	0	0	0	0	0
Calcium Oxide	0	0	0	0	0	0	0	0	0	0	0	0	0	0	0
Benzene	0	0	0	0	0	0	0	0	0	0	0	0	0	0	0
Naphthalene	0	0	0	0	0	0	0	0	0	0	0	0	0	0	0

Table A-1. (Continued)

Stream No.	636	701	702	703	706	707	708	710	711	711	712	712	715	718	719
Hybrid Pop NREL	0	0	0	0	0	0	0	0	0	0	0	0	0	0	0
Sulfur Dioxide	0	0	0	0	0	0	0	0	0	0	0	0	0	0	0
Hydrogen Cyanide	0	0	0	0	0	0	0	0	0	0	0	0	0	0	0
Nitric Oxide	0	0	0	0	0	0	0	0	0	0	0	0	0	0	0
Methanol	0.1456	0	0	0	0	0	0	0	0	0	0	0	0	0	0
Ethanol	15.119	0	0	0	0	0	0	0	0	0	0	0	0	0	0
Isopropanol	0	0	0	0	0	0	0	0	0	0	0	0	0	0	0
N-Propanol	6345.543	0	0	0	0	0	0	0	0	0	0	0	0	0	0
Isobutanol	0	0	0	0	0	0	0	0	0	0	0	0	0	0	0
N-Butanol	2715.64	0	0	0	0	0	0	0	0	0	0	0	0	0	0
1-Pentanol	1354.882	0	0	0	0	0	0	0	0	0	0	0	0	0	0

Stream No.	720	721	725	731	732	737	741	742	743	744	745	746	747	751
Stream Name	FROM STM	FROM SR REC		To STM REF		TO 1945			to GASIFIE		REBOI TO MOL SIEVE			
Temp F	1000	1000	1000	840.662	840.662	526.5776	366.3953	366.3953	366.3953	366.3953	366.3953	366.3953	366.3953	115.5419
Pres psia	850	850	850	450	450	860	35	35	35	35	35	35	35	1.5
Enth MM Btu/h	-2334.2	-678.21	-3012.4	-3053.7	-734.35	-932.53	-2411.1	-413.15	-104.53	-660.77	-152.56	-1076.4	-3.7509	-681.05
Vapor mass fraction	1	1	1	1	1	0.98971	1	1	1	1	1	1	1	0.93746
Total lb/h	435348	126493.8	561842	561842	135113	164146	426729	73120	18500	116945	27000	190500	663.8382	116945

Table A-1. (Continued)

Stream No.	720	721	725	731	732	736	737	741	742	743	744	745	746	747	751
Flowrates in lb/h															
Oxygen	0	0	0	0	0	0	0	0	0	0	0	0	0	0	0
Nitrogen	0	0	0	0	0	0	0	0	0	0	0	0	0	0	0
Argon	0	0	0	0	0	0	0	0	0	0	0	0	0	0	0
Carbon	0	0	0	0	0	0	0	0	0	0	0	0	0	0	0
Hydrogen	0	0	0	0	0	0	0	0	0	0	0	0	0	0	0
Carbon Monoxide	0	0	0	0	0	0	0	0	0	0	0	0	0	0	0
Carbon Dioxide	0	0	0	0	0	0	0	0	0	0	0	0	0	0	0
Methane	0	0	0	0	0	0	0	0	0	0	0	0	0	0	0
Acetylene	0	0	0	0	0	0	0	0	0	0	0	0	0	0	0
Ethylene	0	0	0	0	0	0	0	0	0	0	0	0	0	0	0
Ethane	0	0	0	0	0	0	0	0	0	0	0	0	0	0	0
Propane	0	0	0	0	0	0	0	0	0	0	0	0	0	0	0
Water	435348	126493.8	561842	561842	135113	426729	164146	426729	73120	18500	116945	27000	190500	663.8382	116945
Sulphur	0	0	0	0	0	0	0	0	0	0	0	0	0	0	0
Carbonyl Sulfide	0	0	0	0	0	0	0	0	0	0	0	0	0	0	0
Hydrogen Sulfide	0	0	0	0	0	0	0	0	0	0	0	0	0	0	0
Ammonia	0	0	0	0	0	0	0	0	0	0	0	0	0	0	0
Hydrogen Chloride	0	0	0	0	0	0	0	0	0	0	0	0	0	0	0
Silicon Dioxide	0	0	0	0	0	0	0	0	0	0	0	0	0	0	0
Calcium Oxide	0	0	0	0	0	0	0	0	0	0	0	0	0	0	0
Benzene	0	0	0	0	0	0	0	0	0	0	0	0	0	0	0
Naphthalene	0	0	0	0	0	0	0	0	0	0	0	0	0	0	0

Table A-1. (Continued)

Stream No.	720	721	725	731	732	736	737	741	742	743	744	745	746	747	751
Hybrid Pop NREL	0	0	0	0	0	0	0	0	0	0	0	0	0	0	0
Sulfur Dioxide	0	0	0	0	0	0	0	0	0	0	0	0	0	0	0
Hydrogen Cyanide	0	0	0	0	0	0	0	0	0	0	0	0	0	0	0
Nitric Oxide	0	0	0	0	0	0	0	0	0	0	0	0	0	0	0
Methanol	0	0	0	0	0	0	0	0	0	0	0	0	0	0	0
Ethanol	0	0	0	0	0	0	0	0	0	0	0	0	0	0	0
Isopropanol	0	0	0	0	0	0	0	0	0	0	0	0	0	0	0
N-Propanol	0	0	0	0	0	0	0	0	0	0	0	0	0	0	0
Isobutanol	0	0	0	0	0	0	0	0	0	0	0	0	0	0	0
N-Butanol	0	0	0	0	0	0	0	0	0	0	0	0	0	0	0
1-Pentanol	0	0	0	0	0	0	0	0	0	0	0	0	0	0	0

Stream No.	761	762	769	770	772	773	774	777	779	781	791	792	794	1602	1701
Stream Name			MU from E MOL SIEVEREBOIL COND							BFW			FUEL GASTO WWT		
Temp F	115.5419	115.6498	178.7597	178.7597	259.3462	259.3462	259.3462	226.8287	236.9942	240.2726	526.576	526.5776	1000	71.4942	118.317
Pres psia	1.5	30	60	60	35	35	35	30	30	880	860	860	850	16	15
Enth MM Btu/h	-793.74	-793.72	-1438.1	-1438.1	-179.11	-1264.4	-4.4089	-2922.8	-3821.1	-3819.2	-60.648	-738.52	-2334.2	-171.41	-67250
Vapor mass fraction	0	0	0	0	0.009411	0.00564	0.001	0	0	0	0.99	0.99	1	1	0
Total lb/h	116945	116945	213884	213884	27000	190500	663.8382	437848	573293	573293	10675.7	130000	435348	77146.09	9925413
Flowrates in lb/h															
Oxygen	0	0	0	0	0	0	0	0	0	0	0	0	0	0	0
Nitrogen	0	0	0	0	0	0	0	0	0	0	0	0	0	218.6577	0.0916

Table A-1. (Continued)

Stream No.	761	762	769	770	772	773	774	777	779	781	791	792	794	1602	1701
Argon	0	0	0	0	0	0	0	0	0	0	0	0	0	0	0
Carbon	0	0	0	0	0	0	0	0	0	0	0	0	0	0	0
Hydrogen	0	0	0	0	0	0	0	0	0	0	0	0	0	3434.957	4.1179
Carbon Monoxide	0	0	0	0	0	0	0	0	0	0	0	0	0	44032.44	65.1421
Carbon Dioxide	0	0	0	0	0	0	0	0	0	0	0	0	0	19273.77	17852.19
Methane	0	0	0	0	0	0	0	0	0	0	0	0	0	6844.954	13.669
Acetylene	0	0	0	0	0	0	0	0	0	0	0	0	0	0.0012	0.3676
Ethylene	0	0	0	0	0	0	0	0	0	0	0	0	0	0.2364	13.2064
Ethane	0	0	0	0	0	0	0	0	0	0	0	0	0	223.0892	0.0647
Propane	0	0	0	0	0	0	0	0	0	0	0	0	0	0	0
Water	116945	116945	213884	213884	27000	190500	663.8382	437848	573293	573293	10675.7	130000	435348	126.0698	990037.4
Sulphur	0	0	0	0	0	0	0	0	0	0	0	0	0		0
Carbonyl Sulfide	0	0	0	0	0	0	0	0	0	0	0	0	0		0
Hydrogen Sulfide	0	0	0	0	0	0	0	0	0	0	0	0	0	0	109,948.5
Ammonia	0	0	0	0	0	0	0	0	0	0	0	0	0	10.6638	6979.57
Hydrogen Chloride	0	0	0	0	0	0	0	0	0	0	0	0	0	0	0
Silicon Dioxide	0	0	0	0	0	0	0	0	0	0	0	0	0	0	0
Calcium Oxide	0	0	0	0	0	0	0	0	0	0	0	0	0	0	0
Benzene	0	0	0	0	0	0	0	0	0	0	0	0	0	0	0.2917
Naphthalene	0	0	0	0	0	0	0	0	0	0	0	0	0	0	0.086
Hybrid Pop NREL	0	0	0	0	0	0	0	0	0	0	0	0	0	0	0

Table A-1. (Continued)

Stream No.	761	762	769	770	772	773	774	777	779	781	791	792	794	1602	1701
Sulfur Dioxide	0	0	0	0	0	0	0	0	0	0	0	0	0	0	0
Hydrogen Cyanide	0	0	0	0	0	0	0	0	0	0	0	0	0	0.2293	0
Nitric Oxide	0	0	0	0	0	0	0	0	0	0	0	0	0	0	0
Methanol	0	0	0	0	0	0	0	0	0	0	0	0	0	2293.303	0
Ethanol	0	0	0	0	0	0	0	0	0	0	0	0	0	609.9102	0
Isopropanol	0	0	0	0	0	0	0	0	0	0	0	0	0	0	0
N-Propanol	0	0	0	0	0	0	0	0	0	0	0	0	0	59.8768	0
Isobutanol	0	0	0	0	0	0	0	0	0	0	0	0	0	0	0
N-Butanol	0	0	0	0	0	0	0	0	0	0	0	0	0	12.3067	0
1-Pentanol	0	0	0	0	0	0	0	0	0	0	0	0	0	5.6418	0

Stream No.	1702	1703	1741	1742	1743	1744	1745	1746	1747	1748	1749	1750	1751	1830	1831
Stream Name								TO WWT	TO WWT	to WWT		TO DEARE	to SCRUBBCWS		
Temp F	118.317	118.317	253.0462	274.5935	0	195.4107	195.4109	209.0953	205.1707	231.2065	211.696	171.8389	110	90	110
Pres psia	15	15	28	54	0	109.5	109.5	220	109.5	465	109.5	109.5	427.5	60	60
Enth MM Btu/h	-672.5	-66577	0	0	0	-15.876	-15.876	-39.36	-55.236	-18.357	-73.593	-121.97	-820.73	-26023	-25947
Vapor mass fraction	0	0	0	0	0	0	0	0	0	0	0	0	0	0	0
Total lb/h	99254.12	9826158	0	0	0	2367.09	2367.09	5880.385	8247.476	2751.756	10999.23	18121.07	120823.2	3819701	3819701
Flowrates in lb/h															
Oxygen	0	0	0	0	0	0	0	0	0	0	0	0	0	0	0
Nitrogen	0.0009	0.0907	0	0	0	0	0	0	0	0	0	0	0	0	0
Argon	0	0	0	0	0	0	0	0	0	0	0	0	0	0	0

Table A-1. (Continued)

Stream No.	1702	1703	1741	1742	1743	1744	1745	1746	1747	1748	1749	1750	1751	1830	1831
Carbon	0	0	0	0	0	0	0	0	0	0	0	0	0	0	0
Hydrogen	0.0412	4.0767	0	0	0	0	0	0	0	0	0	0	0	0	0
Carbon Monoxide	0.6514	64.4907	0	0	0	0	0	0	0	0	0	0	0	0	0
Carbon Dioxide	178.5219	17673.67	0	0	0	0	0	0	0	0	0	0	0	0	0
Methane	0.1367	13.5323	0	0	0	0	0	0	0	0	0	0	0	0	0
Acetylene	0.0037	0.3639	0	0	0	0	0	0	0	0	0	0	0	0	0
Ethylene	0.1321	13.0743	0	0	0	0	0	0	0	0	0	0	0	0	0
Ethane	0.0006	0.0641	0	0	0	0	0	0	0	0	0	0	0	0	0
Propane	0	0	0	0	0	0	0	0	0	0	0	0	0	0	0
Water	99003.73	9801370	0	0	0	2367.09	2367.09	5880.385	8247.476	2751.756	10999.23	18121.07	120823.2	3819701	3819701
Sulphur	0	0	0	0	0	0	0	0	0	0	0	0	0	0	0
Carbonyl Sulfide	0	0	0	0	0	0	0	0	0	0	0	0	0	0	0
Hydrogen Sulfide	1.0995	108.849	0	0	0	0	0	0	0	0	0	0	0	0	0
Ammonia	69.7957	6909.775	0	0	0	0	0	0	0	0	0	0	0	0	0
Hydrogen Chloride	0	0	0	0	0	0	0	0	0	0	0	0	0	0	0
Silicon Dioxide	0	0	0	0	0	0	0	0	0	0	0	0	0	0	0
Calcium Oxide	0	0	0	0	0	0	0	0	0	0	0	0	0	0	0
Benzene	0.0029	0.2888	0	0	0	0	0	0	0	0	0	0	0	0	0
Naphthalene	0.0009	0.0851	0	0	0	0	0	0	0	0	0	0	0	0	0
Hybrid Pop NREL	0	0	0	0	0	0	0	0	0	0	0	0	0	0	0

Table A-1. (Continued)

Stream No.	1702	1703	1741	1742	1743	1744	1745	1746	1747	1748	1749	1750	1751	1830	1831
Sulfur Dioxide	0	0	0	0	0	0	0	0	0	0	0	0	0	0	0
Hydrogen Cyanide	0	0	0	0	0	0	0	0	0	0	0	0	0	0	0
Nitric Oxide	0	0	0	0	0	0	0	0	0	0	0	0	0	0	0
Methanol	0	0	0	0	0	0	0	0	0	0	0	0	0	0	0
Ethanol	0	0	0	0	0	0	0	0	0	0	0	0	0	0	0
Isopropanol	0	0	0	0	0	0	0	0	0	0	0	0	0	0	0
N-Propanol	0	0	0	0	0	0	0	0	0	0	0	0	0	0	0
Isobutanol	0	0	0	0	0	0	0	0	0	0	0	0	0	0	0
N-Butanol	0	0	0	0	0	0	0	0	0	0	0	0	0	0	0
1-Pentanol	0	0	0	0	0	0	0	0	0	0	0	0	0	0	0

Stream No.	1834	1835	1840	1841	1842	1843	1850	1851	1861	1862	1871	1872	1910	1931	1932
Stream Name	CWS		CWS		CWS		CWS				STEAM		BFW	to STM DR	
Temp F	90	110	90	110	90	110	90	110	90	110	90	110	260	237	526.5776
Pres psia	60	60	65	65	65	65	65	60	60	60	65	65	25	860	860
Enth MM Btu/h	-196.29	-195.71	-941.89	-939.13	-5031.6	-5016.9	-2485.9	-2478.6	-1183.8	-1179	-62829	-62716	-416.82	-71.155	-60.648
Vapor mass fraction	0	0	0	0	0	0	0	0	0	0	0	0	1	0	0.99
Total lb/h	28811.37	28811.37	138253.4	138253.4	738555	738555	364879	364879	173765	173765	9232755	9232755	73119.59	10675.7	10675.7
Flowrates in lb/h															
Oxygen	0	0	0	0	0	0	0	0	0	0	0	0	0	0	0
Nitrogen	0	0	0	0	0	0	0	0	0	0	0	0	0	0	0

Table A-1. (Continued)

Stream No.	1834	1835	1840	1841	1842	1843	1850	1851	1861	1862	1871	1872	1910	1931	1932
Argon	0	0	0	0	0	0	0	0	0	0	0	0	0	0	0
Carbon	0	0	0	0	0	0	0	0	0	0	0	0	0	0	0
Hydrogen	0	0	0	0	0	0	0	0	0	0	0	0	0	0	0
Carbon Monoxide	0	0	0	0	0	0	0	0	0	0	0	0	0	0	0
Carbon Dioxide	0	0	0	0	0	0	0	0	0	0	0	0	0	0	0
Methane	0	0	0	0	0	0	0	0	0	0	0	0	0	0	0
Acetylene	0	0	0	0	0	0	0	0	0	0	0	0	0	0	0
Ethylene	0	0	0	0	0	0	0	0	0	0	0	0	0	0	0
Ethane	0	0	0	0	0	0	0	0	0	0	0	0	0	0	0
Propane	0	0	0	0	0	0	0	0	0	0	0	0	0	0	0
Water	28811.37	28811.37	138253.4	138253.4	738555	738555	364879	364879	173765	173765	9232755	9232755	73119.59	10675.7	10675.7
Sulphur	0	0	0	0	0	0	0	0	0	0	0	0	0	0	0
Carbonyl Sulfide	0	0	0	0	0	0	0	0	0	0	0	0	0	0	0
Hydrogen Sulfide	0	0	0	0	0	0	0	0	0	0	0	0	0	0	0
Ammonia	0	0	0	0	0	0	0	0	0	0	0	0	0	0	0
Hydrogen Chloride	0	0	0	0	0	0	0	0	0	0	0	0	0	0	0
Silicon Dioxide	0	0	0	0	0	0	0	0	0	0	0	0	0	0	0
Calcium Oxide	0	0	0	0	0	0	0	0	0	0	0	0	0	0	0
Benzene	0	0	0	0	0	0	0	0	0	0	0	0	0	0	0
Naphthalene	0	0	0	0	0	0	0	0	0	0	0	0	0	0	0

Table A-1. (Continued)

Stream No.	1834	1835	1840	1841	1842	1843	1850	1851	1861	1862	1871	1872	1910	1931	1932
Hybrid Pop NREL	0	0	0	0	0	0	0	0	0	0	0	0	0	0	0
Sulfur Dioxide	0	0	0	0	0	0	0	0	0	0	0	0	0	0	0
Hydrogen Cyanide	0	0	0	0	0	0	0	0	0	0	0	0	0	0	0
Nitric Oxide	0	0	0	0	0	0	0	0	0	0	0	0	0	0	0
Methanol	0	0	0	0	0	0	0	0	0	0	0	0	0	0	0
Ethanol	0	0	0	0	0	0	0	0	0	0	0	0	0	0	0
Isopropanol	0	0	0	0	0	0	0	0	0	0	0	0	0	0	0
N-Propanol	0	0	0	0	0	0	0	0	0	0	0	0	0	0	0
Isobutanol	0	0	0	0	0	0	0	0	0	0	0	0	0	0	0
N-Butanol	0	0	0	0	0	0	0	0	0	0	0	0	0	0	0
Stream No.	1935	1936	1938	1940	1941	1942	1943	1944	1945	1946	1950	1951	1952	1953	1954
Stream Name	BFW	MP STEAM	to STM DRSTEAM			STEAM	TO TURBINE		STEAM		BFW	TO STM DBFW			
Temp F	237	400	526.5776	715.0002	525.2153	1000	525.2153	1000	366.3939	259.3462	237	526.5776	237	526.5776	526.5776
Pres psia	860	860	860	450	850	850	850	850	35	35	860	860	860	860	860
Enth MM Btu/h	-866.47	-844.28	-738.52	-743.52	-717.71	-678.21	-2470.1	-2334.2	-3.7505	-4.4085	-1788.5	-1524.3	-1094.1	-932.53	-932.53
Vapor mass fraction	0	0	0.99	1	1	1	1	1	1	0.001	0	0.99	0	0.98971	0.98971
Total lb/h	130000	130000	130000	135112.5	126493.8	126493.8	435348	435348	663.7736	663.7736	268328.3	268328.3	164146	164146	164146
Flowrates in lb/h															
Oxygen	0	0	0	0	0	0	0	0	0	0	0	0	0	0	0

Table A-1. (Continued)

Stream No.	1935	1936	1938	1940	1941	1942	1943	1944	1945	1946	1950	1951	1952	1953	1954
Nitrogen	0	0	0	0	0	0	0	0	0	0	0	0	0	0	0
Argon	0	0	0	0	0	0	0	0	0	0	0	0	0	0	0
Carbon	0	0	0	0	0	0	0	0	0	0	0	0	0	0	0
Hydrogen	0	0	0	0	0	0	0	0	0	0	0	0	0	0	0
Carbon Monoxide	0	0	0	0	0	0	0	0	0	0	0	0	0	0	0
Carbon Dioxide	0	0	0	0	0	0	0	0	0	0	0	0	0	0	0
Methane	0	0	0	0	0	0	0	0	0	0	0	0	0	0	0
Acetylene	0	0	0	0	0	0	0	0	0	0	0	0	0	0	0
Ethylene	0	0	0	0	0	0	0	0	0	0	0	0	0	0	0
Ethane	0	0	0	0	0	0	0	0	0	0	0	0	0	0	0
Propane	0	0	0	0	0	0	0	0	0	0	0	0	0	0	0
Water	130000	130000	130000	135112.5	126493.8	126493.8	435348	435348	663.7736	663.7736	268328.3	268328.3	164146	164146	164146
Sulphur	0	0	0	0	0	0	0	0	0	0	0	0	0	0	0
Carbonyl Sulfide	0	0	0	0	0	0	0	0	0	0	0	0	0	0	0
Hydrogen Sulfide	0	0	0	0	0	0	0	0	0	0	0	0	0	0	0
Ammonia	0	0	0	0	0	0	0	0	0	0	0	0	0	0	0
Hydrogen Chloride	0	0	0	0	0	0	0	0	0	0	0	0	0	0	0
Silicon Dioxide	0	0	0	0	0	0	0	0	0	0	0	0	0	0	0
Calcium Oxide	0	0	0	0	0	0	0	0	0	0	0	0	0	0	0
Benzene	0	0	0	0	0	0	0	0	0	0	0	0	0	0	0
Naphthalene	0	0	0	0	0	0	0	0	0	0	0	0	0	0	0

Table A-1. (Continued)

Stream No.	1935	1936	1938	1940	1941	1942	1943	1944	1945	1946	1950	1951	1952	1953	1954
Hybrid Pop NREL	0	0	0	0	0	0	0	0	0	0	0	0	0	0	0
Sulfur Dioxide	0	0	0	0	0	0	0	0	0	0	0	0	0	0	0
Hydrogen Cyanide	0	0	0	0	0	0	0	0	0	0	0	0	0	0	0
Nitric Oxide	0	0	0	0	0	0	0	0	0	0	0	0	0	0	0
Methanol	0	0	0	0	0	0	0	0	0	0	0	0	0	0	0
Ethanol	0	0	0	0	0	0	0	0	0	0	0	0	0	0	0
Isopropanol	0	0	0	0	0	0	0	0	0	0	0	0	0	0	0
N-Propanol	0	0	0	0	0	0	0	0	0	0	0	0	0	0	0
Isobutanol	0	0	0	0	0	0	0	0	0	0	0	0	0	0	0
N-Butanol	0	0	0	0	0	0	0	0	0	0	0	0	0	0	0
1-Pentanol	0	0	0	0	0	0	0	0	0	0	0	0	0	0	0

Stream No.	1834	1835	1840	1841	1842	1843	1850	1851	1861	1862	1871	1872	1910	1931	1932
Stream Name	CWS		CWS		CWS	CWS	CWS				STEAM		BFW	to STM DR	
Temp F	90	110	90	110	110	90	90	110	90	110	90	110	260	237	526.5776
Pres psia	60	60	65	65	65	65	65	60	60	60	65	65	25	860	860
Enth MM Btu/h	-196.29	-195.71	-941.89	-939.13	-5031.6	-5016.9	-2485.9	-2478.6	-1183.8	-1179	-62829	-62716	-416.82	-71.155	-60.648
Vapor mass fraction	0	0	0	0	0	0	0	0	0	0	0	0	1	0	0.99
Total lb/h	28811.37	28811.37	138253.4	138253.4	738555	738555	364879	364879	173765	173765	9232755	9232755	73119.59	10675.7	10675.7
Flowrates in lb/h															

Table A-1. (Continued)

Stream No.	1834	1835	1840	1841	1842	1843	1850	1851	1861	1862	1871	1872	1910	1931	1932
Oxygen	0	0	0	0	0	0	0	0	0	0	0	0	0	0	0
Nitrogen	0	0	0	0	0	0	0	0	0	0	0	0	0	0	0
Argon	0	0	0	0	0	0	0	0	0	0	0	0	0	0	0
Carbon	0	0	0	0	0	0	0	0	0	0	0	0	0	0	0
Hydrogen	0	0	0	0	0	0	0	0	0	0	0	0	0	0	0
Carbon Monoxide	0	0	0	0	0	0	0	0	0	0	0	0	0	0	0
Carbon Dioxide	0	0	0	0	0	0	0	0	0	0	0	0	0	0	0
Methane	0	0	0	0	0	0	0	0	0	0	0	0	0	0	0
Acetylene	0	0	0	0	0	0	0	0	0	0	0	0	0	0	0
Ethylene	0	0	0	0	0	0	0	0	0	0	0	0	0	0	0
Ethane	0	0	0	0	0	0	0	0	0	0	0	0	0	0	0
Propane	0	0	0	0	0	0	0	0	0	0	0	0	0	0	0
Water	28811.37	28811.37	138253.4	138253.4	738555	738555	364879	364879	173765	173765	9232755	9232755	73119.59	10675.7	10675.7
Sulphur	0	0	0	0	0	0	0	0	0	0	0	0	0	0	0
Carbonyl Sulfide	0	0	0	0	0	0	0	0	0	0	0	0	0	0	0
Hydrogen Sulfide	0	0	0	0	0	0	0	0	0	0	0	0	0	0	0
Ammonia	0	0	0	0	0	0	0	0	0	0	0	0	0	0	0
Hydrogen Chloride	0	0	0	0	0	0	0	0	0	0	0	0	0	0	0
Silicon Dioxide	0	0	0	0	0	0	0	0	0	0	0	0	0	0	0
Calcium Oxide	0	0	0	0	0	0	0	0	0	0	0	0	0	0	0
Benzene	0	0	0	0	0	0	0	0	0	0	0	0	0	0	0

Table A-1. (Continued)

Stream No.	1834	1835	1840	1841	1842	1843	1850	1851	1861	1862	1871	1872	1910	1931	1932
Naphthalene	0	0	0	0	0	0	0	0	0	0	0	0	0	0	0
Hybrid Pop NREL	0	0	0	0	0	0	0	0	0	0	0	0	0	0	0
Sulfur Dioxide	0	0	0	0	0	0	0	0	0	0	0	0	0	0	0
Hydrogen Cyanide	0	0	0	0	0	0	0	0	0	0	0	0	0	0	0
Nitric Oxide	0	0	0	0	0	0	0	0	0	0	0	0	0	0	0
Methanol	0	0	0	0	0	0	0	0	0	0	0	0	0	0	0
Ethanol	0	0	0	0	0	0	0	0	0	0	0	0	0	0	0
Isopropanol	0	0	0	0	0	0	0	0	0	0	0	0	0	0	0
N-Propanol	0	0	0	0	0	0	0	0	0	0	0	0	0	0	0
Isobutanol	0	0	0	0	0	0	0	0	0	0	0	0	0	0	0
N-Butanol	0	0	0	0	0	0	0	0	0	0	0	0	0	0	0
1-Pentanol	0	0	0	0	0	0	0	0	0	0	0	0	0	0	0

APPENDIX B. HEAT AND MATERIAL BALANCE FOR THE MSW CASE

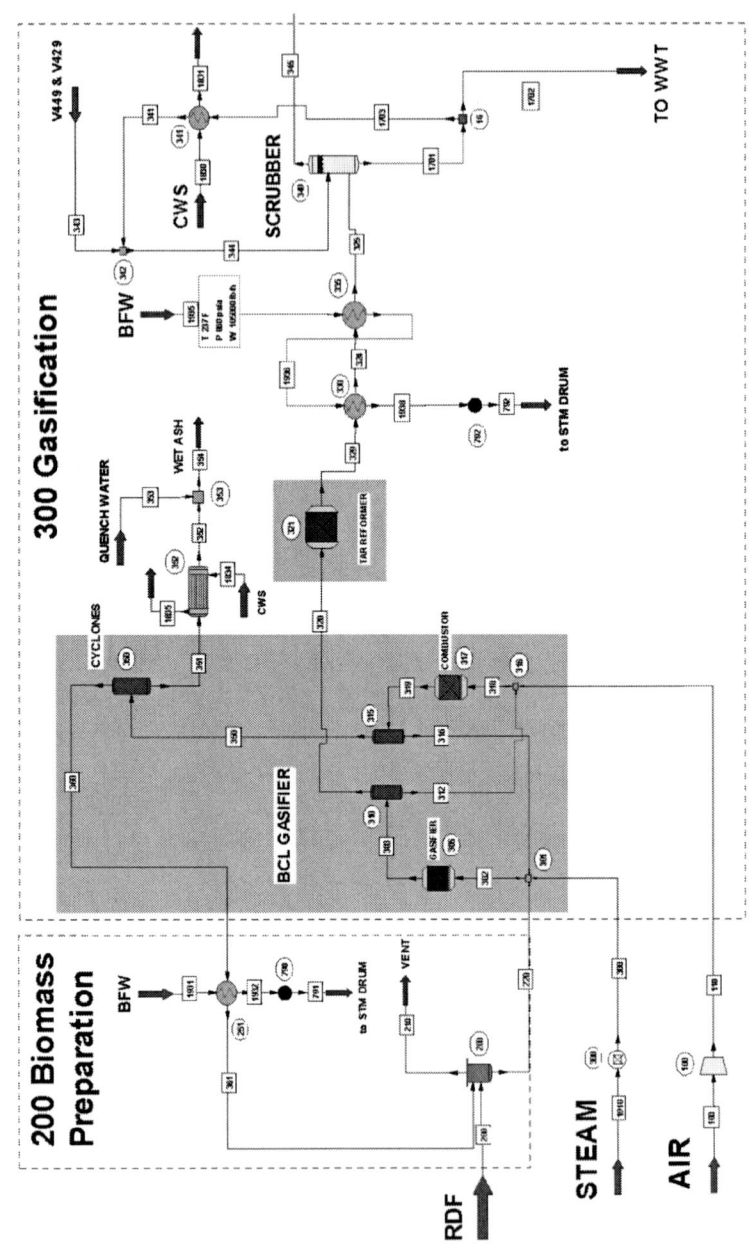

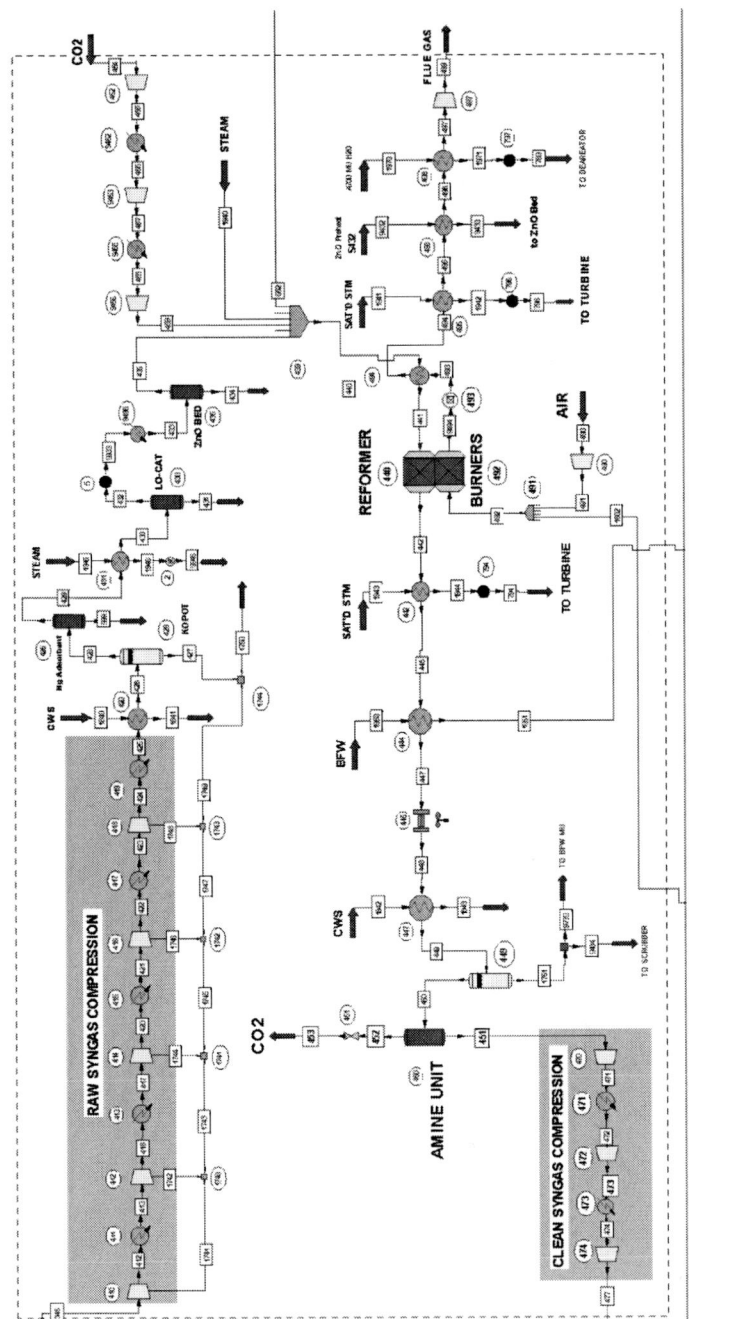

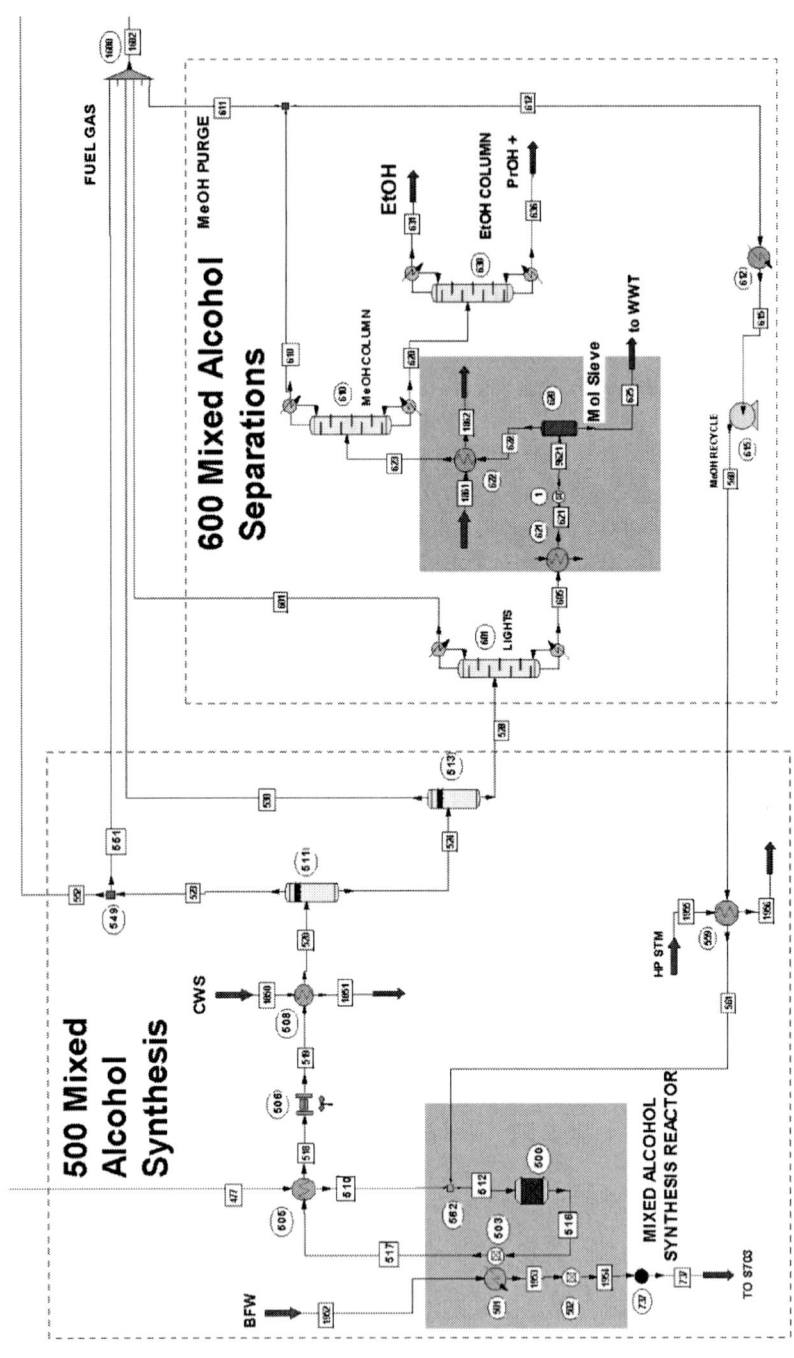

700 STEAM SYSTEM

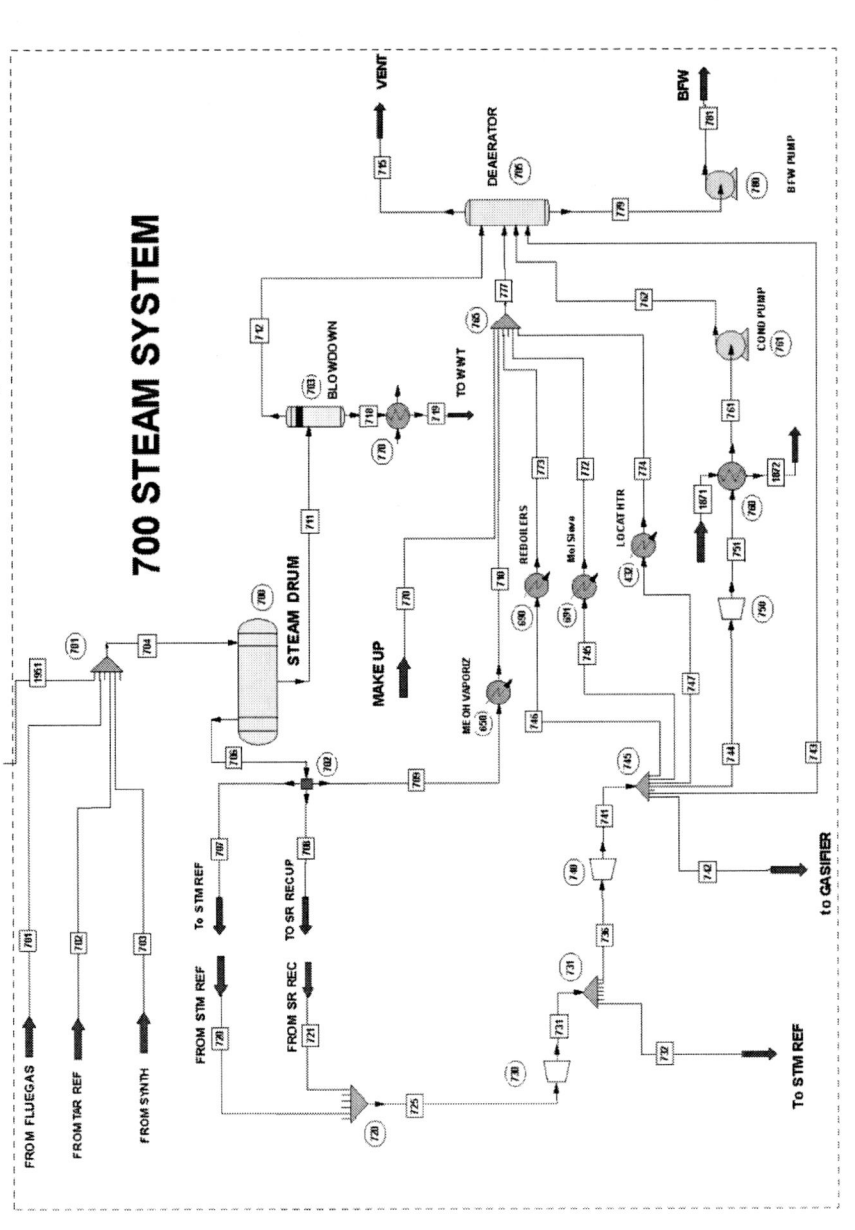

Table B-1. Steam Results for the MSW Case

Stream No.	100	110	200	210	220	300	302	303	312	316	318	319	320	324	325
Stream Name	AIR		RDF	VENT								Syngas			
Temp F	60	172.5231	60	304.7456	220	259.3482	1661.5451	1511	1511	1730	1432.019	1730	1511	462.4696	249.0378
Pres psia	14.696	30	25	23	23	35	23	23	23	23	23	23	23	19	18
Enth MMBtu/h	-71.027	-50.695	-1853	-1746.7	-721.28	-417.08	-11319	-11320	-10777	-10181	-10827	-10828	-543.26	-583.54	-601.47
Vapor mass fraction	0.99137	1	0	1	0	1	1	1	1	0	1	1	1	1	1
Total lb/h	555001	555001	367435.188	795754	204130.703	73119.7656	6277250.5	6277251	6096737	6000000	6651737.5	6651739.5	180513.516	180513.109	180513.109
Flowrates in lb/h															
Oxygen	125921.063	125921.063	0	16851.8438	0	0	0	32906.8516	32906.8516	0	158828	16851.8438	0	0	0
Nitrogen	410968.219	410968.219	0	410945	0	0	0	0	0	0	410968.219	410945	0	392.5347	392.5347
Argon	7009.0005	7009.0005	0	7009.0005	0	0	0	0	0	0	7009.0005	7009.0005	0	0	0
Carbon	0	0	0	0	0	0	0	40123.7344	40123.7344	0	40123.7305	0.0015	0	0	0
Hydrogen	0	0	0	0	0	0	0	5699.1343	4416.0098	0	4416.0098	0.0005	1283.1241	5422.354	5422.354
Carbon Monoxide	0	0	0	0	0	0	0	39389.8164	0	0	0	0	39389.8164	67260.8438	67260.8438
Carbon Dioxide	273.6321	273.6321	0	147297.656	0	0	0	14855.46	0	0	273.6321	147297.656	14855.46	14855.46	14855.46
Methane	0	0	0	0	0	0	0	9054.3818	0	0	0	0	9054.3818	7243.5059	7243.5059

Table B-1. (Continued)

Stream No.	100	110	200	210	220	300	302	303	312	316	318	319	320	324	325
Acetylene	0	0	0	0	0	0	0	398.3862	0	0	0	0	398.3862	199.1931	199.1931
Ethylene	0	0	0	0	0	0	0	17005.373	0	0	0	0	17005.373	8502.6865	8502.6865
Ethane	0	0	0	0	0	0	0	1041.9557	0	0	0	0	1041.9557	104.1956	104.1956
Propane	0	0	0	0	0	0	0	0	0	0	0	0	0	0	0
Water	10828.5596	10828.5596	183717.547	213600	20413.0703	73119.7656	93532.8359	93532.8359	0	0	10828.5596	50295.5	93532.8359	75607.2188	75607.2188
Sulphur	0	0	0	0	0	0	0	0	0	0	0	0	0	0	0
Carbonyl Sulfide	0	0	0	0	0	0	0	0	0	0	0	0	0	0	0
Hydrogen Sulfide	0	0	0	0	0	0	0	397.3538	0	0	0	0	397.3538	397.3538	397.3538
Ammonia	0	0	0	0	0	0	0	681.8284	0	0	0	0	681.8284	204.5485	204.5485
Hydrogen Chloride	0	0	0	0	0	0	0	0	0	0	0	0	0	0	0
Silicon Dioxide	0	0	0	0	0	0	6000000	6000000	6000000	6000000	6000000	6000000	0	0	0
Calcium Oxide	0	0	0	0	0	0	0	19290.3516	19290.3516	0	19290.3496	19290.4082	0	0	0
Benzene	0	0	0	0	0	0	0	718.251	0	0	0	0	718.251	215.4753	215.4753
Naphthalene	0	0	0	0	0	0	0	2154.7529	0	0	0	0	2154.7529	107.7377	107.7377
Hybrid Poplar Ch	0	0	0	0	0	0	0	0	0	0	0	0	0	0	0
Sulfur Dioxide	0	0	0	0	0	0	0	0	0	0	0	0	0	0	0

Table B-1. (Continued)

Stream No.	100	110	200	210	220	300	302	303	312	316	318	319	320	324	325
Hydrogen Cyanide	0	0	0	0	0	0	0	0	0	0	0	0	0	0	0
Nitric Oxide	0	0	0	50.1734	0	0	0	0	0	0	0	50.1734	0	0	0
Methanol	0	0	0	0	0	0	0	0	0	0	0	0	0	0	0
Ethanol	0	0	0	0	0	0	0	0	0	0	0	0	0	0	0
Isopropanol	0	0	0	0	0	0	0	0	0	0	0	0	0	0	0
N-Propanol	0	0	0	0	0	0	0	0	0	0	0	0	0	0	0
Isobutanol	0	0	0	0	0	0	0	0	0	0	0	0	0	0	0
N-Butanol	0	0	0	0	0	0	0	0	0	0	0	0	0	0	0
1-Pentanol	0	0	0	0	0	0	0	0	0	0	0	0	0	0	0
RDF	0	0	183717.625	0	183717.625	0	183717.625	0	0	0	0	0	0	0	0

Stream No.	329	341	343	344	345	350	351	352	353	354	360	361	412	413	416
Stream Name		V449 & V429					ASH		QUENCH WAWET ASH						
Temp F	1383	110	110	109.9994	118.6996	1730	1730	300	60	212.0109	1730	1450	253.7146	140	272.2724
Pres psia	20	15	415	15	15	23	23	23	14.7	14.7	23	23	28	28	54
Enth MMBtu/h	-498.12	-66129	-253.51	-66383	-254.29	-647.04	-86.863	-93.021	-5.1321	-98.153	-560.18	-614.95	-247.14	-254	-246.15
Vapor mass fraction	1	0.0046978	0	0.0046794	1	1	0	0	0	0.33858	1	1	1	0.99327	1
Total lb/h	180513.109	9760903	37319.6602	9798222	118171.734	651740	19290.4082	19290.4082	750.0001	20040.4082	632449	632449	118171.734	118171.734	118171.734
Flowrates in lb/h															
Oxygen	0			0	0	16851.8438	0	0	0	0	16851.8438	16851.8438	0	0	0

Table B-1. (Continued)

Stream No.	329	341	343	344	345	350	351	352	353	354	360	361	412	413	416
Nitrogen	392.5347	0.2354	0	0.2354	392.5323	410945	0	0	0	0	410945	410945	392.5323	392.5323	392.5323
Argon	0	0	0	0	0	7009.0005	0	0	0	0	7009.0005	7009.0005	0	0	0
Carbon	0	0	0	0	0	0.0015	0.0015	0.0015	0	0.0015	0	0	0	0	0
Hydrogen	5422.354	4.7008	0	4.7007	5422.3066	0.0005	0	0	0	0	0.0005	0.0005	5422.3066	5422.3066	5422.3066
Carbon Monoxide	67260.8438	59.6037	0	59.6037	67260.2344	0	0	0	0	0	0	0	67260.2344	67260.2344	67260.2344
Carbon Dioxide	14855.46	29694.6875	0	29694.6914	14548.0752	147297.656	0	0	0	0	147297.656	147297.656	14548.0752	14548.0752	14548.0752
Methane	7243.5059	10.1854	0	10.1854	7243.4019	0	0	0	0	0	0	0	7243.4019	7243.4019	7243.4019
Acetylene	199.1931	0.2868	0	0.2868	199.1902	0	0	0	0	0	0	0	199.1902	199.1902	199.1902
Ethylene	8502.6865	37.3303	0	37.3303	8502.3047	0	0	0	0	0	0	0	8502.3047	8502.3047	8502.3047
Ethane	104.1956	0.1556	0	0.1556	104.194	0	0	0	0	0	0	0	104.194	104.194	104.194
Propane	0	0	0	0	0	0	0	0	0	0	0	0	0	0	0
Water	75607.2188	9716314	373319.6602	9753634	13727.6865	50295.5	0	0	750.0001	750.0001	50295.5	50295.5	13727.6865	13727.6865	13727.6865
Sulphur	0	0	0	0	0	0	0	0	0	0	0	0	0	0	0
Carbonyl Sulfide	0	0	0	0	0	0	0	0	0	0	0	0	0	0	0
Hydrogen Sulfide	397.3538	956.1025	0	956.1025	387.4926	0	0	0	0	0	0	0	387.4926	387.4926	387.4926
Ammonia	204.5485	13825.0391	0	13825.0391	61.1072	0	0	0	0	0	0	0	61.1072	61.1072	61.1072

Table B-1. (Continued)

Stream No.	329	341	343	344	345	350	351	352	353	354	360	361	412	413	416
HydrogenChloride	0	0	0	0	0	0	0	0	0	0	0	0	0	0	0
Silicon Dioxide	0	0	0	0	0	0	0	0	0	0	0	0	0	0	0
Calcium Oxide	0	0	0	0	0	19290.4082	19290.4082	19290.4082	0	19290.4082	0	0	0	0	0
Benzene	215.4753	0.4609	0	0.4609	215.4706	0	0	0	0	0	0	0	215.4706	215.4706	215.4706
Naphthalene	107.7377	0.1397	0	0.1397	107.7363	0	0	0	0	0	0	0	107.7363	107.7363	107.7363
Hybrid Poplar Ch	0	0	0	0	0	0	0	0	0	0	0	0	0	0	0
Sulfur Dioxide	0	0	0	0	0	0	0	0	0	0	0	0	0	0	0
Hydrogen Cyanide	0	0	0	0	0	0	0	0	0	0	0	0	0	0	0
Nitric Oxide	0	0	0	0	0	50.1734	0	0	0	0	50.1734	50.1734	0	0	0
Methanol	0	0	0	0	0	0	0	0	0	0	0	0	0	0	0
Ethanol	0	0	0	0	0	0	0	0	0	0	0	0	0	0	0
Isopropanol	0	0	0	0	0	0	0	0	0	0	0	0	0	0	0
N-Propanol	0	0	0	0	0	0	0	0	0	0	0	0	0	0	0
Isobutanol	0	0	0	0	0	0	0	0	0	0	0	0	0	0	0
N-Butanol	0	0	0	0	0	0	0	0	0	0	0	0	0	0	0
1-Pentanol	0	0	0	0	0	0	0	0	0	0	0	0	0	0	0
RDF	0	0	0	0	0	0	0	0	0	0	0	0	0	0	0
Stream No.	417	420	421	422	423	424	425	426	427	428	429	430	431	432	433
Stream Name															
Temp F	140	195.4259	140	209.108	140	231.2095	140	110	110	110	110	120	120	120	707

Table B-1. (Continued)

Stream No.	417	420	421	422	423	424	425	426	427	428	429	430	431	432	433
Pres psia	54	109.5	109.5	220	220	465	465	465	465	465	455	455	445	445	445
Enth MMBtu/h	-260.68	-239.67	-251.46	-214.38	-223.94	-201.82	-211.25	-213.28	-35.361	-177.92	-177.91	-177.43	-0.10025	-177.36	-148.05
Vapor mass fraction	0.9376	1	0.92429	1	0.94641	1	0.95643	0.95175	0	1	1	1	0	1	1
Total lb/h	118171.734	116309.711	116309.711	112006.781	112006.781	109992.469	109992.469	109992.469	5306.8062	104685.656	104685.656	104685.656	61.0737	104624.586	104624.586
Flowrates in lb/h															
Oxygen	0	0	0	0	0	0	0	0	0	0	0	0	0	0	0
Nitrogen	392.5323	392.5323	392.5323	392.5323	392.5323	392.5323	392.5323	392.5323	0.0018	392.5304	392.5304	392.5304	0	392.5304	392.5304
Argon	0	0	0	0	0	0	0	0	0	0	0	0	0	0	0
Carbon	0	0	0	0	0	0	0	0	0	0	0	0	0	0	0
Hydrogen	5422.3066	5422.3066	5422.3066	5422.3066	5422.3066	5422.3066	5422.3066	5422.3066	0.0108	5422.2959	5422.2959	5422.2959	0	5422.2959	5422.2959
Carbon Monoxide	67260.2344	67260.2344	67260.2344	67260.2344	67260.2344	67260.2344	67260.2344	67260.2344	0.3159	67259.9141	67259.9141	67259.9141	0	67259.9141	67259.9141
Carbon Dioxide	14548.0752	14548.0742	14548.0742	14548.0723	14548.0723	14548.0723	14548.0723	14548.0723	1.7083	14546.3652	14546.3652	14546.3652	0	14546.3652	14546.3652
Methane	7243.4019	7243.4019	7243.4019	7243.4019	7243.4019	7243.4019	7243.4019	7243.4019	0.1385	7243.2627	7243.2627	7243.2627	0	7243.2627	7243.2627
Acetylene	199.1902	199.1902	199.1902	199.1902	199.1902	199.1902	199.1902	199.1902	0.0265	199.1637	199.1637	199.1637	0	199.1637	199.1637
Ethylene	8502.3047	8502.3047	8502.3047	8502.3037	8502.3037	8502.3037	8502.3037	8502.3037	0.703	8501.6016	8501.6016	8501.6016	0	8501.6016	8501.6016
Ethane	104.194	104.194	104.194	104.194	104.194	104.194	104.194	104.194	0.0125	104.1815	104.1815	104.1815	0	104.1815	104.1815
Propane	0	0	0	0	0	0	0	0	0	0	0	0	0	0	0

Table B-1. (Continued)

Stream No.	417	420	421	422	423	424	425	426	427	428	429	430	431	432	433
Water	13727.6865	11865.6709	11865.6709	7562.7437	7562.7437	5548.4297	5548.4297	5548.4297	5209.1787	339.2523	339.2523	339.2523	0	339.2523	339.2523
Sulphur	0	0	0	0	0	0	0	0	0	0	0	0	0	0	0
Carbonyl Sulfide	0	0	0	0	0	0	0	0	0	0	0	0	0	0	0
Hydrogen Sulfide	387.4926	387.4926	387.4926	387.4925	387.4925	387.4925	387.4925	387.4925	0.1328	387.3598	387.3598	387.3598	0	387.3598	387.3598
Ammonia	61.1072	61.1072	61.1072	61.1072	61.1072	61.1072	61.1072	61.1072	0.0335	61.0737	61.0737	61.0737	61.0737	0	0
HydrogenChloride	0	0	0	0	0	0	0	0	0	0	0	0	0	0	0
Silicon Dioxide	0	0	0	0	0	0	0	0	0	0	0	0	0	0	0
Calcium Oxide	0	0	0	0	0	0	0	0	0	0	0	0	0	0	0
Benzene	215.4706	215.4706	215.4706	215.4706	215.4706	215.4706	215.4706	215.4706	5.2171	210.2535	210.2535	210.2535	0	210.2535	210.2535
Naphthalene	107.7363	107.7363	107.7363	107.7363	107.7363	107.7363	107.7363	107.7363	89.3262	18.4101	18.4101	18.4101	0	18.4101	18.4101
Hybrid Poplar Ch	0	0	0	0	0	0	0	0	0	0	0	0	0	0	0
Sulfur Dioxide	0	0	0	0	0	0	0	0	0	0	0	0	0	0	0
Hydrogen Cyanide	0	0	0	0	0	0	0	0	0	0	0	0	0	0	0
Nitric Oxide	0	0	0	0	0	0	0	0	0	0	0	0	0	0	0
Methanol	0	0	0	0	0	0	0	0	0	0	0	0	0	0	0
Ethanol	0	0	0	0	0	0	0	0	0	0	0	0	0	0	0
Isopropanol	0	0	0	0	0	0	0	0	0	0	0	0	0	0	0
N-Propanol	0	0	0	0	0	0	0	0	0	0	0	0	0	0	0

Table B-1. (Continued)

Stream No.	417	420	421	422	423	424	425	426	427	428	429	430	431	432	433
Isobutanol	0	0	0	0	0	0	0	0	0	0	0	0	0	0	0
N-Butanol	0	0	0	0	0	0	0	0	0	0	0	0	0	0	0
1-Pentanol	0	0	0	0	0	0	0	0	0	0	0	0	0	0	0
RDF	0	0	0	0	0	0	0	0	0	0	0	0	0	0	0
Stream No.	434	435	440	441	442	445	447	448	449	450	451	452	453	454	455
Stream Name												CO2		CO2	
Temp F	707	707	468.973 2	900	1652	1155.37 76	300.007 4	150	110	110	120	120	57.2347	60	316.152 6
Pres psia	682	682	450	445	430	430	430	430	427.5	427.5	422.5	422.5	22	22	100
Enth MMBtu/h	-0.04066	-147.95	-1847	-1752.4	-1485.8	-1606.4	-1853.9	-1950.4	-1963.4	-1116.4	-318.22	-798.47	-798.47	-762.12	-751.25
Vapor mass fraction	1	1	1	1	1	1	0.88387	0.77007	0.76623	1	1	0.99634	0.99895	1	1
Total lb/h	387.359 8	104237. 227	533428	533428	533428	533428	533428	533428	533428	408726. 344	201888. 734	206837. 609	206837. 609	198045	198045
Flowrates in lb/h															
Oxygen	0	0	0	0	0	0	0	0	0	0	0	0	0	0	0
Nitrogen	0	392.530 4	965.587 7	965.587 7	956.474 9	956.474 9	956.474 9	956.474 9	956.474 9	956.474 9	956.474 9	0	0	0	0
Argon	0	0	0	0	0	0	0	0	0	0	0	0	0	0	0
Carbon	0	0	0	0	0	0	0	0	0	0	0	0	0	0	0
Hydrogen	0	5422.29 59	10511.9 482	10511.9 482	14833.0 586	14833.0 586	14833.0 586	14833.0 586	14833.0 586	14833.0 576	14833.0 576	0	0	0	0
Carbon Monoxide	0	67259.9 141	128707. 758	128707. 758	176995. 125	176995. 125	176995. 125	176995. 125	176995. 125	176995. 125	176995. 125	0	0	0	0
Carbon Dioxide	0	14546.3 652	233360. 594	233360. 594	207796. 297	207796. 297	207796. 297	207796. 297	207796. 297	207796. 297	2077.96 83	205718. 328	205718. 328	198045	198045
Methane	0	7243.26 27	14122.8 584	14122.8 584	7008.36 82	7008.36 82	7008.36 82	7008.36 82	7008.36 82	7008.36 82	7008.36 82	0	0	0	0

Table B-1. (Continued)

Stream No.	434	435	440	441	442	445	447	448	449	450	451	452	453	454	455
Acetylene	0	199.1637	199.1642	199.1642	0.0011	0.0011	0.0011	0.0011	0.0011	0.0011	0.0011	0	0	0	0
Ethylene	0	8501.6016	8501.7246	8501.7246	0.2139	0.2139	0.2139	0.2139	0.2139	0.2139	0.2139	0	0	0	0
Ethane	0	104.1815	376.0543	376.0543	0.4389	0.4389	0.4389	0.4389	0.4389	0.4389	0.4389	0	0	0	0
Propane	0	0	0	0	0	0	0	0	0	0	0	0	0	0	0
Water	0	339.2523	135551.422	135551.422	125820.68	125820.68	125820.68	125820.68	125820.68	1119.2845	0	1119.2845	1119.2869	0	0
Sulphur	0	0	0	0	0	0	0	0	0	0	0	0	0	0	0
Carbonyl Sulfide	0	0	0	0	0	0	0	0	0	0	0	0	0	0	0
Hydrogen Sulfide	387.3598	0	0	0	0	0	0	0	0	0	0	0	0	0	0
Ammonia	0	0	5.8146	5.8146	16.797	16.797	16.797	16.797	16.797	16.797	16.797	0	0	0	0
HydrogenChloride	0	0	0	0	0	0	0	0	0	0	0	0	0	0	0
Silicon Dioxide	0	0	0	0	0	0	0	0	0	0	0	0	0	0	0
Calcium Oxide	0	0	0	0	0	0	0	0	0	0	0	0	0	0	0
Benzene	0	210.2535	210.2535	210.2535	0	0	0	0	0	0	0	0	0	0	0
Naphthalene	0	18.4101	18.4101	18.4101	0	0	0	0	0	0	0	0	0	0	0
Hybrid Poplar Ch	0	0	0	0	0	0	0	0	0	0	0	0	0	0	0
Sulfur Dioxide	0	0	0	0	0	0	0	0	0	0	0	0	0	0	0
Hydrogen Cyanide	0	0	0.0371	0.0371	0.2021	0.2021	0.2021	0.2021	0.2021	0.2021	0.2021	0	0	0	0

Table B-1. (Continued)

Stream No.	434	435	440	441	442	445	447	448	449	450	451	452	453	454	455
Nitric Oxide	0	0	0	0	0	0	0	0	0	0	0	0	0	0	0
Methanol	0	0	316.2478	316.2478	0.0746	0.0746	0.0746	0.0746	0.0746	0.0746	0.0746	0	0	0	0
Ethanol	0	0	506.4705	506.4705	0.0001	0.0001	0.0001	0.0001	0.0001	0.0001	0.0001	0	0	0	0
Isopropanol	0	0	0	0	0	0	0	0	0	0	0	0	0	0	0
N-Propanol	0	0	55.838	55.838	0	0	0	0	0	0	0	0	0	0	0
Isobutanol	0	0	0	0	0	0	0	0	0	0	0	0	0	0	0
N-Butanol	0	0	12.14	12.14	0	0	0	0	0	0	0	0	0	0	0
1-Pentanol	0	0	5.6754	5.6754	0	0	0	0	0	0	0	0	0	0	0
RDF	0	0	0	0	0	0	0	0	0	0	0	0	0	0	0
Stream No.	456	457	458	459	471	472	473	474	477	490	491	492	493	494	495
Stream Name										AIR			To LOCAT&ZnO		
Temp F	150	317.6958	150	256.3385	236.1337	150	278.5218	150	271.5315	60	124.5558	112.3454	1800	1099.1305	753.4954
Pres psia	100	250	250	450	700	700	1200	1200	2000	14.696	20	16	16	14	14
Enth MMBtu/h	-758.75	-751.58	-759.43	-755.14	-306.39	-315.21	-301.9	-315.24	-302.35	-1.522	4.135	-148.4	-415.2	-509.72	-553.33
Vapor mass fraction	1	1	1	1	1	1	1	1	1	1	1	1	1	1	1
Total lb/h	198045	198045	198045	198045	2018887.34	2018887.34	2018887.34	2018887.34	2018887.34	362548.563	362548.563	432050	432050	432050	432050
Flowrates in lb/h															
Oxygen	0	0	0	0	0	0	0	0	0	84442.6797	84442.6797	84442.6797	0	0	0
Nitrogen	0	0	0	0	956.4749	956.4749	956.4749	956.4749	956.4749	278106	278106	278489.313	11010.9072	11010.9072	11010.9072
Argon	0	0	0	0	0	0	0	0	0	0	0	0	0	0	0
Carbon	0	0	0	0	0	0	0	0	0	0	0	0	0	0	0

Table B-1. (Continued)

Stream No.	456	457	458	459	471	472	473	474	477	490	491	492	493	494	495
Hydrogen	0	0	0	0	14833.0576	14833.0576	14833.0576	14833.0576	14833.0576	0	0	3405.0361	0.0012	0.0012	0.0012
Carbon Monoxide	0	0	0	0	176995.125	176995.125	176995.125	176995.125	176995.125	0	0	41200.3203	0.0246	0.0246	0.0246
Carbon Dioxide	198045	198045	198045	198045	2077.9683	2077.9683	2077.9683	2077.9683	2077.9683	0	0	17461.6094	98743.8594	98743.8594	98743.8594
Methane	0	0	0	0	7008.3682	7008.3682	7008.3682	7008.3682	7008.3682	0	0	4690.6763	0.0003	0.0003	0.0003
Acetylene	0	0	0	0	0.0011	0.0011	0.0011	0.0011	0.0011	0	0	0.0005	0.0004	0.0004	0.0004
Ethylene	0	0	0	0	0.2139	0.2139	0.2139	0.2139	0.2139	0	0	0.0904	0.0004	0.0004	0.0004
Ethane	0	0	0	0	0.4389	0.4389	0.4389	0.4389	0.4389	0	0	203.5968	0.0004	0.0004	0.0004
Propane	0	0	0	0	0	0	0	0	0	0	0	0	0	0	0
Water	0	0	0	0	0	0	0	0	0	0	0	66.4548	43773.25	43773.25	43773.25
Sulphur	0	0	0	0	0	0	0	0	0	0	0	0	0	0	0
Carbonyl Sulfide	0	0	0	0	0	0	0	0	0	0	0	0	0	0	0
Hydrogen Sulfide	0	0	0	0	0	0	0	0	0	0	0	0	0	0	0
Ammonia	0	0	0	0	16.797	16.797	16.797	16.797	16.797	0	0	10.9824	0.0003	0.0003	0.0003
HydrogenChloride	0	0	0	0	0	0	0	0	0	0	0	0	0	0	0
Silicon Dioxide	0	0	0	0	0	0	0	0	0	0	0	0	0	0	0
Calcium Oxide	0	0	0	0	0	0	0	0	0	0	0	0	0	0	0
Benzene	0	0	0	0	0	0	0	0	0	0	0	0	0	0	0
Naphthalene	0	0	0	0	0	0	0	0	0	0	0	0	0	0	0

Table B-1. (Continued)

Stream No.	456	457	458	459	471	472	473	474	477	490	491	492	493	494	495
Hybrid Poplar Ch	0	0	0	0	0	0	0	0	0	0	0	0	0	0	0
Hydrogen Cyanide	0	0	0	0	0.2021	0.2021	0.2021	0.2021	0.2021	0	0	0.165	0.0004	0.0004	0.0004
Nitric Oxide	0	0	0	0	0	0	0	0	0	0	0	0	44.0383	44.0383	44.0383
Methanol	0	0	0	0	0.0746	0.0746	0.0746	0.0746	0.0746	0	0	1686.7837	0.0005	0.0005	0.0005
Ethanol	0	0	0	0	0.0001	0.0001	0.0001	0.0001	0.0001	0	0	342.8955	0.0007	0.0007	0.0007
Isopropanol	0	0	0	0	0	0	0	0	0	0	0	0	0	0	0
N-Propanol	0	0	0	0	0	0	0	0	0	0	0	37.2493	0.0008	0.0008	0.0008
Isobutanol	0	0	0	0	0	0	0	0	0	0	0	0	0	0	0
N-Butanol	0	0	0	0	0	0	0	0	0	0	0	8.0933	0.001	0.001	0.001
1-Pentanol	0	0	0	0	0	0	0	0	0	0	0	3.7836	0.0011	0.0011	0.0011
RDF	0	0	0	0	0	0	0	0	0	0	0	0	0	0	0
Stream No.	496	497	499	510	512	516	517	518	519	520	523	524	528	530	551
Stream Name	FLUE GAS					GAS OUT					TO WWT				
Temp F	510.6381	280	297.8182	450	451.0915	570	570	384.1223	150	110	110.2502	110.2502	110.2502	110.2502	110.2502
Pres psia	14	14	15	1995	1995	1995	1995	1990	1980	1975	1975	1975	1975	1975	1975
Enth MMBtu/h	-582.63	-609.5	-607.45	-283.68	-308.14	-429.72	-429.72	-448.39	-487.36	-492.79	-333.89	-158.9	-158.9	0	-133.56
Vapor mass fraction	1	1	1	1	1	1	1	1	0.77735	0.76051	1	0	0	1	1
Total lb/h	432050	432050	432050	201888.734	210582.625	210584.063	210584.063	210584.063	210584.063	210584.063	160055.422	50528.6328	50528.6328	0	64022.1836
Flowrates in lb/h															

Table B-1. (Continued)

Stream No.	456	457	458	459	471	472	473	474	477	490	491	492	493	494	495
Oxygen	11010.9072	11010.9072	11010.9072	0	0	0	0	0	0	0	0	0	0	0	0
Nitrogen	278478	278478	278478	956.475	956.4749	956.4749	956.4749	956.4749	956.4749	956.4749	955.0955	1.3794	1.3794	0	382.0382
Argon	0	0	0	0	0	0	0	0	0	0	0	0	0	0	0
Carbon	0	0	0	0	0	0	0	0	0	0	0	0	0	0	0
Hydrogen	0.0012	0.0012	0.0012	14833.0586	14833.0576	8494.6875	8494.6875	8494.6875	8494.6875	8494.6875	8482.7539	11.9344	11.9344	0	3393.1018
Carbon Monoxide	0.0246	0.0246	0.0246	176995.125	176995.094	102648.141	102648.141	102648.141	102648.141	102648.141	102413.047	235.0997	235.0997	0	40965.2188
Carbon Dioxide	98743.8594	98743.8594	98743.8594	2077.9683	2077.9683	38230.8398	38230.8398	38230.8398	38230.8398	38230.8398	34615.3867	3615.4543	3615.4543	0	13846.1553
Methane	0.0003	0.0003	0.0003	7008.3687	7008.3662	11570.2715	11570.2715	11570.2715	11570.2715	11570.2715	11466	104.2797	104.2797	0	4586.3965
Acetylene	0.0004	0.0004	0.0004	0.0011	0.0011	0.0011	0.0011	0.0011	0.0011	0.0011	0.001	0.0001	0.0001	0	0.0004
Ethylene	0.0004	0.0004	0.0004	0.2139	0.2139	0.2139	0.2139	0.2139	0.2139	0.2139	0.2058	0.0081	0.0081	0	0.0823
Ethane	0.0004	0.0004	0.0004	0.4389	0.4389	475.4695	475.4695	475.4695	475.4695	475.4695	453.1212	22.3483	22.3483	0	181.2485
Propane	0	0	0	0		0	0	0	0	0	0	0	0	0	0
Water	43773.25	43773.25	43773.25	0		5526.8784	5526.8784	5526.8784	5526.8784	5526.8784	166.1309	5360.7461	5360.7461	0	66.4524
Sulphur	0	0	0	0	0	0	0	0	0	0	0	0	0	0	0
Carbonyl Sulfide	0	0	0	0	0	0	0	0	0	0	0	0	0	0	0
Hydrogen Sulfide	0	0	0	0	0	0	0	0	0	0	0	0	0	0	0
Ammonia	0.0003	0.0003	0.0003	16.797	16.797	16.797	16.797	16.797	16.797	16.797	9.691	7.1059	7.1059	0	3.8764
HydrogenChloride	0	0	0	0	0	0	0	0	0	0	0	0	0	0	0

Table B-1. (Continued)

Stream No.	456	457	458	459	471	472	473	474	477	490	491	492	493	494	495
Silicon Dioxide	0	0	0	0	0	0	0	0	0	0	0	0	0	0	0
Calcium Oxide	0	0	44.0383	0	0	0	0	0	0	0	0	0	0	0	0
Benzene	0	0	0	0	0	0	0	0	0	0	0	0	0	0	0
Naphthalene	0	0	0	0	0	0	0	0	0	0	0	0	0	0	0
Hybrid Poplar Ch	0	0	0	0	0	0	0	0	0	0	0	0	0	0	0
Sulfur Dioxide	0	0	0	0	0	0	0	0	0	0	0	0	0	0	0
Hydrogen Cyanide	0.0004	0.0004	0.0004	0.2021	0.2021	0.2021	0.2021	0.2021	0.2021	0.2021	0.0618	0.1403	0.1403	0	0.0247
Nitric Oxide	44.0383	44.0383	0.0005	0	0	0	0	0	0	0	0	0	0	0	0
Methanol	0.0005	0.0005	0.0005	0.0746	8672.9873	10773.2061	10773.2061	10773.2061	10773.2061	10773.2061	527.0796	10246.125	10246.125	0	210.8318
Ethanol	0.0007	0.0007	0.0007	0.0001	20.8168	23846.5332	23846.5332	23846.5332	23846.5332	23846.5332	844.1176	23002.4082	23002.4082	0	337.647
Isopropanol	0	0	0	0	0	0	0	0	0	0	0	0	0	0	0
N-Propanol	0.0008	0.0008	0.0008	0	0.2151	4936.3354	4936.3354	4936.3354	4936.3354	4936.3354	93.0634	4843.271	4843.271	0	37.2254
Isobutanol	0	0	0	0	0	0	0	0	0	0	0	0	0	0	0
N-Butanol	0.001	0.001	0.001	0	0	2073.8054	2073.8054	2073.8054	2073.8054	2073.8054	20.2333	2053.5718	2053.5718	0	8.0933
1-Pentanol	0.0011	0.0011	0.0011	0	0	1034.2206	1034.2206	1034.2206	1034.2206	1034.2206	9.459	1024.7612	1024.7612	0	3.7836
RDF	0	0	0	0	0	0	0	0	0	0	0	0	0	0	0
Stream No.	552	560	561	601	605	610	611	612	615	620	621	622	623	625	631
Stream Name	MeOH RECYCLE	MeOH RECYCLE				MeOH	MeOH PURG	MeOH RECYCLE	RECYCLE		Mixed OH		to WWT		EtOH
Temp F	110.2502	167.2084	480	84.2883	194.8814	152.0141	152.0141	152.0141	152.0141	197.947	201.8105	201.8105	187.3736	201.8105	176.7085

Table B-1. (Continued)

Stream No.	552	560	561	601	605	610	611	612	615	620	621	622	623	625	631
Pres psia	1975	2000	1995	23	26.7	16	16	16	16	22	26.7	23	23	23	16
Enth MMBtu/h	-200.33	-27.353	-24.466	-15.934	-139.72	-30.493	-3.0493	-27.444	-27.444	-73.487	-119.57	-93.378	-103.99	-35.881	-58.387
Vapor mass fraction	1	0.0003125	1	1	0	0	0	0	0	0	1	0.67884	1.75E-06	0	0
Total lb/h	96033.2734	8693.9443	8693.9443	4512.9829	46015.6484	9659.9404	965.994	8693.9443	8693.9443	30995.0215	46015.6484	40655	40655	5360.7534	23110.4844
Flowrates in lb/h															
Oxygen	0	0	0	0	0	0	0	0	0	0	0	0	0	0	0
Nitrogen	573.0573	0	0	1.3794	0	0	0	0	0	0	0	0	0	0	0
Argon	0	0	0	0	0	0	0	0	0	0	0	0	0	0	0
Carbon	0	0	0	0	0	0	0	0	0	0	0	0	0	0	0
Hydrogen	5089.6528	0	0	11.9344	0	0	0	0	0	0	0	0	0	0	0
Carbon Monoxide	61447.832	0	0	235.0997	0	0	0	0	0	0	0	0	0	0	0
Carbon Dioxide	20769.2324	0	0	3615.4543	0	0	0	0	0	0	0	0	0	0	0
Methane	6879.5952	0	0	104.2797	0	0	0	0	0	0	0	0	0	0	0
Acetylene	0.0006	0	0	0.0001	0	0	0	0	0	0	0	0	0	0	0
Ethylene	0.1235	0	0	0.0081	0	0	0	0	0	0	0	0	0	0	0
Ethane	271.8728	0	0	22.3483	0	0	0	0	0	0	0	0	0	0	0
Propane	0	0	0	0	0	0	0	0	0	0	0	0	0	0	0
Water	99.6786	0	0	0.0025	5360.7437	0	0	0	0	0	5360.7437	0	0	5360.7534	0
Sulphur	0	0	0	0	0	0	0	0	0	0	0	0	0	0	0

Table B-1. (Continued)

Stream No.	552	560	561	601	605	610	611	612	615	620	621	622	623	625	631
Carbonyl Sulfide	0	0	0	0	0	0	0	0	0	0	0	0	0	0	0
Hydrogen Sulfide	0	0	0	0	0	0	0	0	0	0	0	0	0	0	0
Ammonia	5.8146	0	0	7.1059	0	0	0	0	0	0	0	0	0	0	0
HydrogenChloride	0	0	0	0	0	0	0	0	0	0	0	0	0	0	0
Silicon Dioxide	0	0	0	0	0	0	0	0	0	0	0	0	0	0	0
Calcium Oxide	0	0	0	0	0	0	0	0	0	0	0	0	0	0	0
Benzene	0	0	0	0	0	0	0	0	0	0	0	0	0	0	0
Naphthalene	0	0	0	0	0	0	0	0	0	0	0	0	0	0	0
Hybrid Poplar Ch	0	0	0	0	0	0	0	0	0	0	0	0	0	0	0
Sulfur Dioxide	0.0371	0	0	0	0	0	0	0	0	0	0	0	0	0	0
Hydrogen Cyanide	0	0	0	0.1403	0	0	0	0	0	0	0	0	0	0	0
Nitric Oxide	0	0	0	0	0	0	0	0	0	0	0	0	0	0	0
Methanol	316.2478	8672.9131	8672.9131	512.2948	9733.8291	9636.5713	963.6571	8672.9131	8672.9131	97.2767	9733.8291	9733.8477	9733.8477	0	97.166
Ethanol	506.4705	20.8167	20.8167	2.9355	22999.4727	23.1297	2.313	20.8167	20.8167	22976.373	22999.4727	22999.502	22999.502	0	22964.8906
Isopropanol	0	0	0	0	0	0	0	0	0	0	0	0	0	0	0
N-Propanol	55.838	0.2151	0.2151	0	4843.271	0.239	0.0239	0.2151	0.2151	4843.0371	4843.271	4843.2759	4843.2759	0	48.4271
Isobutanol	0	0	0	0	0	0	0	0	0	0	0	0	0	0	0
N-Butanol	12.14	0	0	0	2053.5718	0	0	0	0	2053.5737	2053.5718	2053.5737	2053.5737	0	0

Table B-1. (Continued)

Stream No.	552	560	561	601	605	610	611	612	615	620	621	622	623	625	631
1-Pentanol	5.6754	0	0	0	1024.7612	0	0	0	0	1024.7623	1024.7612	1024.7623	1024.7623	0	0
RDF	0	0	0	0	0	0	0	0	0	0	0	0	0	0	0
Stream No.	636	701	702	703	704	706	707	708	709	710	711	712	715	718	719
Stream Name	PrOH +	FROM FLUEG	FROM TAR R	FROM SYNTH			To STM REF TO SR RECUP						VENT	TO WWT	
Temp F	230.6059	526.5776	526.5776	526.5776	526.5776	526.5776	526.5776	526.5776	526.5776	526.5776	526.5776	526.5776	233.9624	526.5776	150
Pres psia	19	860	860	860	860	860	860	860	860	860	860	860	30	860	860
Enth MMBtu/h	-15.264	-316.11	-596.49	-701.44	-3042.5	-3008.8	-2191.7	-792.39	-24.683	-27.583	-33.679	-3.96E-05	0	-33.679	-35.83
Vapor mass fraction	0	0.99	0.99	0.99043	0.99009	1	1	1	1	0.0095721	0	1	1	0	0
Total lb/h	7884.5376	55644.2734	105000	123479.414	535571	530265	386267	139648	4350.0005	4350.0005	5305.981	0.007	0	5305.9736	5305.9736
Flowrates in lb/h															
Oxygen	0	0	0	0	0	0	0	0	0	0	0	0	0	0	0
Nitrogen	0	0	0	0	0	0	0	0	0	0	0	0	0	0	0
Argon	0	0	0	0	0	0	0	0	0	0	0	0	0	0	0
Carbon	0	0	0	0	0	0	0	0	0	0	0	0	0	0	0
Hydrogen	0	0	0	0	0	0	0	0	0	0	0	0	0	0	0
Carbon Monoxide	0	0	0	0	0	0	0	0	0	0	0	0	0	0	0
Carbon Dioxide	0	0	0	0	0	0	0	0	0	0	0	0	0	0	0
Methane	0	0	0	0	0	0	0	0	0	0	0	0	0	0	0
Acetylene	0	0	0	0	0	0	0	0	0	0	0	0	0	0	0
Ethylene	0	0	0	0	0	0	0	0	0	0	0	0	0	0	0

Table B-1. (Continued)

Stream No.	636	701	702	703	704	706	707	708	709	710	711	712	715	718	719
Ethane	0	0	0	0	0	0	0	0	0	0	0	0	0	0	0
Propane	0	0	0	0	0	0	0	0	0	0	0	0	0	0	0
Water	0	55644.2734	105000	123479.414	535571	530265	386267	139648	4350.0005	4350.0005	5305.981	0.007	0	5305.9736	5305.9736
Sulphur	0	0	0	0	0	0	0	0	0	0	0	0	0	0	0
Carbonyl Sulfide	0	0	0	0	0	0	0	0	0	0	0	0	0	0	0
Hydrogen Sulfide	0	0	0	0	0	0	0	0	0	0	0	0	0	0	0
Ammonia	0	0	0	0	0	0	0	0	0	0	0	0	0	0	0
HydrogenChloride	0	0	0	0	0	0	0	0	0	0	0	0	0	0	0
Silicon Dioxide	0	0	0	0	0	0	0	0	0	0	0	0	0	0	0
Calcium Oxide	0	0	0	0	0	0	0	0	0	0	0	0	0	0	0
Benzene	0	0	0	0	0	0	0	0	0	0	0	0	0	0	0
Naphthalene	0	0	0	0	0	0	0	0	0	0	0	0	0	0	0
Hybrid Poplar Ch	0	0	0	0	0	0	0	0	0	0	0	0	0	0	0
Sulfur Dioxide	0	0	0	0	0	0	0	0	0	0	0	0	0	0	0
Hydrogen Cyanide	0	0	0	0	0	0	0	0	0	0	0	0	0	0	0
Nitric Oxide	0	0	0	0	0	0	0	0	0	0	0	0	0	0	0
Methanol	0.1108	0	0	0	0	0	0	0	0	0	0	0	0	0	0
Ethanol	11.4814	0	0	0	0	0	0	0	0	0	0	0	0	0	0
Isopropanol	0	0	0	0	0	0	0	0	0	0	0	0	0	0	0

Table B-1. (Continued)

Stream No.	636	701	702	703	704	706	707	708	709	710	711	712	715	718	719
N-Propanol	4794.6099	0	0	0	0	0	0	0	0	0	0	0	0	0	0
Isobutanol	0	0	0	0	0	0	0	0	0	0	0	0	0	0	0
N-Butanol	2053.5737	0	0	0	0	0	0	0	0	0	0	0	0	0	0
1-Pentanol	1024.7623	0	0	0	0	0	0	0	0	0	0	0	0	0	0
Stream No.	636	701	702	703	704	706	707	708	709	710	711	712	713	718	719
Stream Name	PrOH +	FROM FLUEG	FROM TAR R	FROM SYNTH			To STM REF	TO SR RECUP					VENT	TO WWT	
Temp F	230.6059	526.5776	526.5776	526.5776	526.5776	526.5776	526.5776	526.5776	526.5776	526.5776	526.5776	526.5776	233.9624	526.5776	150
Pres psia	19	860	860	860	860	860	860	860	860	860	860	860	30	860	860
Enth MMBtu/h	-15.264	-316.11	-596.49	-701.44	-3042.5	-3008.8	-2191.7	-792.39	-24.683	-27.583	-33.679	-3.96E-05	0	-33.679	-35.83
Vapor mass fraction	0	0.99	0.99	0.99043	0.99009	1	1	1	1	0.0095721	0	1	1	0	0
Total lb/h	7884.5376	55644.2734	105000	123479.414	535571	530265	386267	139648	4350.0005	4350.0005	5305.981	0.007	0	5305.9736	5305.9736
Flowrates in lb/h															
Oxygen	0	0	0	0	0	0	0	0	0	0	0	0	0	0	0
Nitrogen	0	0	0	0	0	0	0	0	0	0	0	0	0	0	0
Argon	0	0	0	0	0	0	0	0	0	0	0	0	0	0	0
Carbon	0	0	0	0	0	0	0	0	0	0	0	0	0	0	0
Hydrogen	0	0	0	0	0	0	0	0	0	0	0	0	0	0	0
Carbon Monoxide	0	0	0	0	0	0	0	0	0	0	0	0	0	0	0
Carbon Dioxide	0	0	0	0	0	0	0	0	0	0	0	0	0	0	0

Table B-1. (Continued)

Stream No.	636	701	702	703	704	706	707	708	709	710	711	712	713	718	719
Methane	0	0	0	0	0	0	0	0	0	0	0	0	0	0	0
Acetylene	0	0	0	0	0	0	0	0	0	0	0	0	0	0	0
Ethylene	0	0	0	0	0	0	0	0	0	0	0	0	0	0	0
Ethane	0	0	0	0	0	0	0	0	0	0	0	0	0	0	0
Propane	0	0	0	0	0	0	0	0	0	0	0	0	0	0	0
Water	0	55644.2734	105000	123479.414	535571	530265	386267	139648	4350.0005	4350.0005	5305.981	0.007	0	5305.9736	5305.9736
Sulphur	0	0	0	0	0	0	0	0	0	0	0	0	0	0	0
Carbonyl Sulfide	0	0	0	0	0	0	0	0	0	0	0	0	0	0	0
Hydrogen Sulfide	0	0	0	0	0	0	0	0	0	0	0	0	0	0	0
Ammonia	0	0	0	0	0	0	0	0	0	0	0	0	0	0	0
HydrogenChloride	0	0	0	0	0	0	0	0	0	0	0	0	0	0	0
Silicon Dioxide	0	0	0	0	0	0	0	0	0	0	0	0	0	0	0
Calcium Oxide	0	0	0	0	0	0	0	0	0	0	0	0	0	0	0
Benzene	0	0	0	0	0	0	0	0	0	0	0	0	0	0	0
Naphthalene	0	0	0	0	0	0	0	0	0	0	0	0	0	0	0
Hybrid Poplar Ch	0	0	0	0	0	0	0	0	0	0	0	0	0	0	0
Sulfur Dioxide	0	0	0	0	0	0	0	0	0	0	0	0	0	0	0
Hydrogen Cyanide	0	0	0	0	0	0	0	0	0	0	0	0	0	0	0
Nitric Oxide	0	0	0	0	0	0	0	0	0	0	0	0	0	0	0
Methanol	0.1108	0	0	0	0	0	0	0	0	0	0	0	0	0	0

Table B-1. (Continued)

Stream No.	636	701	702	703	704	706	707	708	709	710	711	712	713	718	719
Ethanol	11.4814	0	0	0	0	0	0	0	0	0	0	0	0	0	0
Isopropanol	0	0	0	0	0	0	0	0	0	0	0	0	0	0	0
N-Propanol	4794.6099	0	0	0	0	0	0	0	0	0	0	0	0	0	0
Isobutanol	0	0	0	0	0	0	0	0	0	0	0	0	0	0	0
N-Butanol	2053.5737	0	0	0	0	0	0	0	0	0	0	0	0	0	0
1-Pentanol	1024.7623	0	0	0	0	0	0	0	0	0	0	0	0	0	0
RDF	0	0	0	0	0	0	0	0	0	0	0	0	0	0	0
Stream No	720	721	725	731	732	736	737	741	742	743	744	745	746	747	751
Stream Name	FROM STM R	FROM SR REC		To STM REF		TO S703		to GASIFIER		TO MOL SIEVE					
Temp F	1000	1000	1000	840.6619	840.6619	840.6619	526.5776	366.397	366.397	366.397	366.397	366.397	366.397	366.397	115.5419
Pres psia	850	850	850	450	450	450	860	35	35	35	35	35	35	35	1.5
Enth MMBtu/h	-2071	-748.74	-2819.8	-2858.4	-734.35	-2124.1	-701.44	-2208.2	-413.15	-94.359	-754.33	-129.96	-813.64	-2.719	-777.48
Vapor mass fraction	1	1	1	1	1	1	0.99043	1	1	1	1	1	1	1	0.93747
Total lb/h	386270.031	139648.406	525918.438	525918.438	135113	390805.438	123479.414	390805.438	73120	16700	133504.219	23000	144000	481.211	133504.219
Flowrates in lb/h															
Oxygen	0	0	0	0	0	0	0	0	0	0	0	0	0	0	0
Nitrogen	0	0	0	0	0	0	0	0	0	0	0	0	0	0	0
Argon	0	0	0	0	0	0	0	0	0	0	0	0	0	0	0
Carbon	0	0	0	0	0	0	0	0	0	0	0	0	0	0	0
Hydrogen	0	0	0	0	0	0	0	0	0	0	0	0	0	0	0

Table B-1. (Continued)

Stream No	720	721	725	731	732	736	737	741	742	743	744	745	746	747	751
Carbon Monoxide	0	0	0	0	0	0	0	0	0	0	0	0	0	0	0
Carbon Dioxide	0	0	0	0	0	0	0	0	0	0	0	0	0	0	0
Methane	0	0	0	0	0	0	0	0	0	0	0	0	0	0	0
Acetylene	0	0	0	0	0	0	0	0	0	0	0	0	0	0	0
Ethylene	0	0	0	0	0	0	0	0	0	0	0	0	0	0	0
Ethane	0	0	0	0	0	0	0	0	0	0	0	0	0	0	0
Propane	0	0	0	0	0	0	0	0	0	0	0	0	0	0	0
Water	386270.031	139648.406	525918.438	525918.438	135113	390805.438	123479.414	390805.438	73120	16700	133504.219	23000	144000	481.211	133504.219
Sulphur	0	0	0	0	0	0	0	0	0	0	0	0	0	0	0
Carbonyl Sulfide	0	0	0	0	0	0	0	0	0	0	0	0	0	0	0
Hydrogen Sulfide	0	0	0	0	0	0	0	0	0	0	0	0	0	0	0
Ammonia	0	0	0	0	0	0	0	0	0	0	0	0	0	0	0
HydrogenChloride	0	0	0	0	0	0	0	0	0	0	0	0	0	0	0
Silicon Dioxide	0	0	0	0	0	0	0	0	0	0	0	0	0	0	0
Calcium Oxide	0	0	0	0	0	0	0	0	0	0	0	0	0	0	0
Benzene	0	0	0	0	0	0	0	0	0	0	0	0	0	0	0
Naphthalene	0	0	0	0	0	0	0	0	0	0	0	0	0	0	0
Hybrid Poplar Ch	0	0	0	0	0	0	0	0	0	0	0	0	0	0	0

Table B-1. (Continued)

Stream No	720	721	725	731	732	736	737	741	742	743	744	745	746	747	751
Sulfur Dioxide	0	0	0	0	0	0	0	0	0	0	0	0	0	0	0
Hydrogen Cyanide	0	0	0	0	0	0	0	0	0	0	0	0	0	0	0
Nitric Oxide	0	0	0	0	0	0	0	0	0	0	0	0	0	0	0
Methanol	0	0	0	0	0	0	0	0	0	0	0	0	0	0	0
Ethanol	0	0	0	0	0	0	0	0	0	0	0	0	0	0	0
Isopropanol	0	0	0	0	0	0	0	0	0	0	0	0	0	0	0
N-Propanol	0	0	0	0	0	0	0	0	0	0	0	0	0	0	0
Isobutanol	0	0	0	0	0	0	0	0	0	0	0	0	0	0	0
N-Butanol	0	0	0	0	0	0	0	0	0	0	0	0	0	0	0
1-Pentanol	0	0	0	0	0	0	0	0	0	0	0	0	0	0	0
RDF	0	0	0	0	0	0	0	0	0	0	0	0	0	0	0
Stream No.	761	762	769	770	772	773	774	777	779	781	791	792	794	796	999
Stream Name		TO DEAREATMAKE UP				MOL SIEVE REBOIL COND				BFW					
Temp F	115.5419	115.6686	185.6402	185.6402	250.3853	259.3462	250.3853	231.0757	233.9624	237.2776	526.5776	526.5776	1000	1000	110
Pres psia	1.5	35	60	60	30	35	30	30	30	890	860	860	850	850	455
Enth MMBtu/h	-906.13	-906.11	-1434.3	-1434.3	-150.11	-955.64	-3.2004	-2570.9	-3571.3	-3569.5	-316.11	-596.49	-2071	-748.74	0
Vapor mass fraction	0	0	0	0	0.13252	0.006452	0.001	0	0	0	0.99	0.99	1	1	0
Total lb/h	133504.219	133504.219	213536	213536	23000	144000	481.211	385367.219	535572	535572	55644.2734	105000	386270.031	139648.406	0
Flowrates in lb/h															
Oxygen	0	0	0	0	0	0	0	0	0	0	0	0	0	0	0
Nitrogen	0	0	0	0	0	0	0	0	0	0	0	0	0	0	0
Argon	0	0	0	0	0	0	0	0	0	0	0	0	0	0	0

Table B-1. (Continued)

Stream No.	761	762	769	770	772	773	774	777	779	781	791	792	794	796	999
Carbon	0	0	0	0	0	0	0	0	0	0	0	0	0	0	0
Hydrogen	0	0	0	0	0	0	0	0	0	0	0	0	0	0	0
Carbon Monoxide	0	0	0	0	0	0	0	0	0	0	0	0	0	0	0
Carbon Dioxide	0	0	0	0	0	0	0	0	0	0	0	0	0	0	0
Methane	0	0	0	0	0	0	0	0	0	0	0	0	0	0	0
Acetylene	0	0	0	0	0	0	0	0	0	0	0	0	0	0	0
Ethylene	0	0	0	0	0	0	0	0	0	0	0	0	0	0	0
Ethane	0	0	0	0	0	0	0	0	0	0	0	0	0	0	0
Propane	0	0	0	0	0	0	0	0	0	0	0	0	0	0	0
Water	133504.219	133504.219	213536	213536	23000	144000	481.211	385367.219	535572	535572	55644.2734	105000	386270.031	139648.406	0
Sulphur	0	0	0	0	0	0	0	0	0	0	0	0	0	0	0
Carbonyl Sulfide	0	0	0	0	0	0	0	0	0	0	0	0	0	0	0
Hydrogen Sulfide	0	0	0	0	0	0	0	0	0	0	0	0	0	0	0
Ammonia	0	0	0	0	0	0	0	0	0	0	0	0	0	0	0
HydrogenChloride	0	0	0	0	0	0	0	0	0	0	0	0	0	0	0
Silicon Dioxide	0	0	0	0	0	0	0	0	0	0	0	0	0	0	0
Calcium Oxide	0	0	0	0	0	0	0	0	0	0	0	0	0	0	0
Benzene	0	0	0	0	0	0	0	0	0	0	0	0	0	0	0
Naphthalene	0	0	0	0	0	0	0	0	0	0	0	0	0	0	0
Hybrid Poplar Ch	0	0	0	0	0	0	0	0	0	0	0	0	0	0	0

Table B-1. (Continued)

Stream No.	761	762	769	770	772	773	774	777	779	781	791	792	794	796	999
Sulfur Dioxide	0	0	0	0	0	0	0	0	0	0	0	0	0	0	0
Hydrogen Cyanide	0	0	0	0	0	0	0	0	0	0	0	0	0	0	0
Nitric Oxide	0	0	0	0	0	0	0	0	0	0	0	0	0	0	0
Methanol	0	0	0	0	0	0	0	0	0	0	0	0	0	0	0
Ethanol	0	0	0	0	0	0	0	0	0	0	0	0	0	0	0
Isopropanol	0	0	0	0	0	0	0	0	0	0	0	0	0	0	0
N-Propanol	0	0	0	0	0	0	0	0	0	0	0	0	0	0	0
Isobutanol	0	0	0	0	0	0	0	0	0	0	0	0	0	0	0
N-Butanol	0	0	0	0	0	0	0	0	0	0	0	0	0	0	0
1-Pentanol	0	0	0	0	0	0	0	0	0	0	0	0	0	0	0
RDF	0	0	0	0	0	0	0	0	0	0	0	0	0	0	0
Stream No.	1602	1701	1702	1703	1741	1742	1743	1744	1745	1746	1747	1748	1749	1750	1751
Stream Name	FUEL GAS TO WWT							TO WWT		to WWT			TO DEAREA to SCRUBBE		
Temp F	75.4944	118.6996	118.7001	118.7001	253.7146	272.2724		195.4259	195.4258	209.108	204.9788	231.2095	211.4502	171.9244	110
Pres psia	16	15	15	15	28	54		109.5	109.5	220	109.5	465	109.5	109.5	427.5
Enth MMBtu/h	-152.54	-66730	-667.23	-66055	0	0		-12.489	-12.489	-28.801	-41.29	-13.438	-54.728	-90.088	-847.07
Vapor mass fraction	1	0	0	0	0	0		0	0	0	0	0	0	0	0
Total lb/h	69501.1563	9860562	98595	9760903	0	0		1862.0194	1862.0194	4302.9282	6164.9473	2014.3109	8179.2583	13486.0625	124701.398
Flowrates in lb/h															
Oxygen	0	0	0	0	0	0	0	0	0	0	0	0	0	0	0
Nitrogen	383.4176	0.2378	0.0024	0.2354	0	0	0	0	0	0	0	0	0	0.0018	0

Table B-1. (Continued)

Stream No.	1602	1701	1702	1703	1741	1742	1743	1744	1745	1746	1747	1748	1749	1750	1751
Argon	0	0	0	0	0	0	0	0	0	0	0	0	0	0	0
Carbon	0	0	0	0	0	0	0	0	0	0	0	0	0	0	0
Hydrogen	3405.0361	4.7488	0.0475	4.7008	0	0	0	0	0	0	0	0	0	0.0108	0
Carbon Monoxide	41200.3203	60.2127	0.6021	59.6037	0	0	0	0	0	0	0	0	0	0.3159	0
Carbon Dioxide	17461.6094	30002.0762	299.9464	29694.6875	0	0	0	0	0	0	0	0	0	1.7083	0
Methane	4690.6763	10.2895	0.1029	10.1854	0	0	0	0	0	0	0	0	0	0.1385	0
Acetylene	0.0005	0.2898	0.0029	0.2868	0	0	0	0	0	0	0	0	0	0.0265	0
Ethylene	0.0904	37.7117	0.3771	37.3303	0	0	0	0	0	0	0	0	0	0.703	0
Ethane	203.5968	0.1572	0.0016	0.1556	0	0	0	0	0	0	0	0	0	0.0125	0
Propane	0	0	0	0	0	0	0	0	0	0	0	0	0	0	0
Water	66.4548	9815512	98144.5938	9716314	0	0	0	1862.0194	1862.0194	4302.9282	6164.9473	2014.3109	8179.2583	13388.4365	124701.398
Sulphur	0	0	0	0	0	0	0	0	0	0	0	0	0	0	0
Carbonyl Sulfide	0	0	0	0	0	0	0	0	0	0	0	0	0	0	0
Hydrogen Sulfide	0	965.9637	9.6576	956.1025	0	0	0	0	0	0	0	0	0	0.1328	0
Ammonia	10.9824	13968.4805	139.6469	13825.0391	0	0	0	0	0	0	0	0	0	0.0335	0
HydrogenChloride	0	0	0	0	0	0	0	0	0	0	0	0	0	0	0
Silicon Dioxide	0	0	0	0	0	0	0	0	0	0	0	0	0	0	0
Calcium Oxide	0	0	0	0	0	0	0	0	0	0	0	0	0	0	0

Table B-1. (Continued)

Stream No.	1602	1701	1702	1703	1741	1742	1743	1744	1745	1746	1747	1748	1749	1750	1751
Benzene	0	0.4656	0.0047	0.4609	0	0	0	0	0	0	0	0	0	5.2171	0
Naphthalene	0	0.1412	0.0014	0.1397	0	0	0	0	0	0	0	0	0	89.3262	0
Hybrid Poplar Ch	0	0	0	0	0	0	0	0	0	0	0	0	0	0	0
Sulfur Dioxide	0	0	0	0	0	0	0	0	0	0	0	0	0	0	0
Hydrogen Cyanide	0.165	0	0	0	0	0	0	0	0	0	0	0	0	0	0
Nitric Oxide	0	0	0	0	0	0	0	0	0	0	0	0	0	0	0
Methanol	1686.7837	0	0	0	0	0	0	0	0	0	0	0	0	0	0
Ethanol	342.8955	0	0	0	0	0	0	0	0	0	0	0	0	0	0
Isopropanol	0	0	0	0	0	0	0	0	0	0	0	0	0	0	0
N-Propanol	37.2493	0	0	0	0	0	0	0	0	0	0	0	0	0	0
Isobutanol	0	0	0	0	0	0	0	0	0	0	0	0	0	0	0
N-Butanol	8.0933	0	0	0	0	0	0	0	0	0	0	0	0	0	0
1-Pentanol	3.7836	0	0	0	0	0	0	0	0	0	0	0	0	0	0
RDF	0	0	0	0	0	0	0	0	0	0	0	0	0	0	0

Stream No.	1830	1831	1834	1835	1840	1841	1842	1843	1850	1851	1861	1862	1871	1872	1910
Stream Name	CWS		CWS		CWS		CWS		CWS						STEAM
Temp F	90	110	90	110	90	110	90	110	90	110	90	110	90	110	259.3482
Pres psia	60	60	60	60	65	65	65	65	65	60	60	60	65	60	35
Enth MMBtu/h	-25113	-25039	-2097.5	-2091.3	-690.94	-688.92	-4435.2	-4422.2	-1849.9	-1844.5	-2604.6	-2594	-43820	-43691	-417.08
Vapor mass fraction	0	0	0	0	0	0	0	0	0	0	0	0	0	0	1

Table B-1. (Continued)

Stream No.	1830	1831	1834	1835	1840	1841	1842	1843	1850	1851	1861	1862	1871	1872	1910
Total lb/h	3686166	3686166	307872	307872	101418.43	101418.43	651012.188	651012.188	271534	271534	382314	382314	6431974	6431974	73119.7656
Flowrates in lb/h															
Oxygen	0	0	0	0	0	0	0	0	0	0	0	0	0	0	0
Nitrogen	0	0	0	0	0	0	0	0	0	0	0	0	0	0	0
Argon	0	0	0	0	0	0	0	0	0	0	0	0	0	0	0
Carbon	0	0	0	0	0	0	0	0	0	0	0	0	0	0	0
Hydrogen	0	0	0	0	0	0	0	0	0	0	0	0	0	0	0
Carbon Monoxide	0	0	0	0	0	0	0	0	0	0	0	0	0	0	0
Carbon Dioxide	0	0	0	0	0	0	0	0	0	0	0	0	0	0	0
Methane	0	0	0	0	0	0	0	0	0	0	0	0	0	0	0
Acetylene	0	0	0	0	0	0	0	0	0	0	0	0	0	0	0
Ethylene	0	0	0	0	0	0	0	0	0	0	0	0	0	0	0
Ethane	0	0	0	0	0	0	0	0	0	0	0	0	0	0	0
Propane	0	0	0	0	0	0	0	0	0	0	0	0	0	0	0
Water	3686166	3686166	307872	307872	101418.43	101418.43	651012.188	651012.188	271534	271534	382314	382314	6431974	6431974	73119.7656
Sulphur	0	0	0	0	0	0	0	0	0	0	0	0	0	0	0
Carbonyl Sulfide	0	0	0	0	0	0	0	0	0	0	0	0	0	0	0
Hydrogen Sulfide	0	0	0	0	0	0	0	0	0	0	0	0	0	0	0
Ammonia	0	0	0	0	0	0	0	0	0	0	0	0	0	0	0
HydrogenChloride	0	0	0	0	0	0	0	0	0	0	0	0	0	0	0

Table B-1. (Continued)

Stream No.	1830	1831	1834	1835	1840	1841	1842	1843	1850	1851	1861	1862	1871	1872	1910
Silicon Dioxide	0	0	0	0	0	0	0	0	0	0	0	0	0	0	0
Calcium Oxide	0	0	0	0	0	0	0	0	0	0	0	0	0	0	0
Benzene	0	0	0	0	0	0	0	0	0	0	0	0	0	0	0
Naphthalene	0	0	0	0	0	0	0	0	0	0	0	0	0	0	0
Hybrid Poplar Ch	0	0	0	0	0	0	0	0	0	0	0	0	0	0	0
Sulfur Dioxide	0	0	0	0	0	0	0	0	0	0	0	0	0	0	0
Hydrogen Cyanide	0	0	0	0	0	0	0	0	0	0	0	0	0	0	0
Nitric Oxide	0	0	0	0	0	0	0	0	0	0	0	0	0	0	0
Methanol	0	0	0	0	0	0	0	0	0	0	0	0	0	0	0
Ethanol	0	0	0	0	0	0	0	0	0	0	0	0	0	0	0
Isopropanol	0	0	0	0	0	0	0	0	0	0	0	0	0	0	0
N-Propanol	0	0	0	0	0	0	0	0	0	0	0	0	0	0	0
Isobutanol	0	0	0	0	0	0	0	0	0	0	0	0	0	0	0
N-Butanol	0	0	0	0	0	0	0	0	0	0	0	0	0	0	0
1-Pentanol	0	0	0	0	0	0	0	0	0	0	0	0	0	0	0
RDF	0	0	0	0	0	0	0	0	0	0	0	0	0	0	0

Stream No.	1931	1932	1935	1936	1935	1940	1941	1942	1943	1944	1945	1946	1950	1951	1952
Stream Name	BFW	to STM DRUMBFW		MP STEAM to STM DRUMSTEAM		STEAM to STM	SATD STM TO TURBINE		STM TO TURBINE	STEAM		to TURBINE BFW		TO STM DRUBFW	
Temp F	237	526.5776	237	400	526.5776	715.0002	525.2153	1000	525.2153	1000	366.401	259.3462	237	526.5776	237
Pres psia	860	860	860	860	860	450	850	850	850	850	35	35	860	860	860
Enth MMBtu/h	-370.88	-316.11	-699.84	-681.92	-596.49	-743.52	-792.35	-748.74	-2191.6	-2071	-2.7189	-3.1959	-1675.9	-1428.4	-823.01

Table B-1. (Continued)

Stream No.	1931	1932	1935	1936	1935	1940	1941	1942	1943	1944	1945	1946	1950	1951	1952
Vapor mass fraction	0	0.99	0	0	0.99	1	1	1	1	1	1	0.001	0	0.99	0
Total lb/h	55644.2734	55644.2734	105000	105000	105000	135112.5	139648.406	139648.406	386270.031	386270.031	481.1948	481.1948	251447	251447	123479.414
Flowrates in lb/h															
Oxygen	0	0	0	0	0	0	0	0	0	0	0	0	0	0	0
Nitrogen	0	0	0	0	0	0	0	0	0	0	0	0	0	0	0
Argon	0	0	0	0	0	0	0	0	0	0	0	0	0	0	0
Carbon	0	0	0	0	0	0	0	0	0	0	0	0	0	0	0
Hydrogen	0	0	0	0	0	0	0	0	0	0	0	0	0	0	0
Carbon Monoxide	0	0	0	0	0	0	0	0	0	0	0	0	0	0	0
Carbon Dioxide	0	0	0	0	0	0	0	0	0	0	0	0	0	0	0
Methane	0	0	0	0	0	0	0	0	0	0	0	0	0	0	0
Acetylene	0	0	0	0	0	0	0	0	0	0	0	0	0	0	0
Ethylene	0	0	0	0	0	0	0	0	0	0	0	0	0	0	0
Ethane	0	0	0	0	0	0	0	0	0	0	0	0	0	0	0
Propane	0	0	0	0	0	0	0	0	0	0	0	0	0	0	0
Water	55644.2734	55644.2734	105000	105000	105000	135112.5	139648.406	139648.406	386270.031	386270.031	481.1948	481.1948	251447	251447	123479.414
Sulphur	0	0	0	0	0	0	0	0	0	0	0	0	0	0	0
Carbonyl Sulfide	0	0	0	0	0	0	0	0	0	0	0	0	0	0	0
Hydrogen Sulfide	0	0	0	0	0	0	0	0	0	0	0	0	0	0	0
Ammonia	0	0	0	0	0	0	0	0	0	0	0	0	0	0	0

Table B-1. (Continued)

Stream No.	1931	1932	1935	1936	1935	1940	1941	1942	1943	1944	1945	1946	1950	1951	1952
HydrogenChloride	0	0	0	0	0	0	0	0	0	0	0	0	0	0	0
Silicon Dioxide	0	0	0	0	0	0	0	0	0	0	0	0	0	0	0
Calcium Oxide	0	0	0	0	0	0	0	0	0	0	0	0	0	0	0
Benzene	0	0	0	0	0	0	0	0	0	0	0	0	0	0	0
Naphthalene	0	0	0	0	0	0	0	0	0	0	0	0	0	0	0
Hybrid Poplar Ch	0	0	0	0	0	0	0	0	0	0	0	0	0	0	0
Sulfur Dioxide	0	0	0	0	0	0	0	0	0	0	0	0	0	0	0
Hydrogen Cyanide	0	0	0	0	0	0	0	0	0	0	0	0	0	0	0
Nitric Oxide	0	0	0	0	0	0	0	0	0	0	0	0	0	0	0
Methanol	0	0	0	0	0	0	0	0	0	0	0	0	0	0	0
Ethanol	0	0	0	0	0	0	0	0	0	0	0	0	0	0	0
Isopropanol	0	0	0	0	0	0	0	0	0	0	0	0	0	0	0
N-Propanol	0	0	0	0	0	0	0	0	0	0	0	0	0	0	0
Isobutanol	0	0	0	0	0	0	0	0	0	0	0	0	0	0	0
N-Butanol	0	0	0	0	0	0	0	0	0	0	0	0	0	0	0
1-Pentanol	0	0	0	0	0	0	0	0	0	0	0	0	0	0	0
RDF	0	0	0	0	0	0	0	0	0	0	0	0	0	0	0
Stream No.	1931	1932	1935	1936	1938	1940	1941	1942	1943	1944	1945	1946	1950	1951	1952
Stream Name	BFW	to STM DRUMBFW	DRUMBFW	MP STEAM to STM DRUMSTEAM			SAT'D STM TO TURBINE				STEAM	to TURBINE BFW		TO STM DRUBFW	
Temp F	237	526.5776	237	400	526.5776	715.0002	525.2153	1000	525.2153	1000	366.401	259.3462	237	526.5776	237
Pres psia	860	860	860	860	860	450	850	850	850	850	35	35	860	860	860

Table B-1. (Continued)

Stream No.	1931	1932	1935	1936	1938	1940	1941	1942	1943	1944	1945	1946	1950	1951	1952
Enth MMBtu/h	-370.88	-316.11	-699.84	-681.92	-596.49	-743.52	-792.35	-748.74	-2191.6	-2071	-2.7189	-3.1959	-1675.9	-1428.4	-823.01
Vapor mass fraction	0	0.99	0	0	0.99	1	1	1	1	1	1	0.001	0	0.99	0
Total lb/h	55644.2734	55644.2734	105000	105000	105000	135112.5	139648.406	139648.406	386270.031	386270.031	481.1948	481.1948	251447	251447	123479.414
Flowrates in lb/h															
Oxygen	0	0	0	0	0	0	0	0	0	0	0	0	0	0	0
Nitrogen	0	0	0	0	0	0	0	0	0	0	0	0	0	0	0
Argon	0	0	0	0	0	0	0	0	0	0	0	0	0	0	0
Carbon	0	0	0	0	0	0	0	0	0	0	0	0	0	0	0
Hydrogen	0	0	0	0	0	0	0	0	0	0	0	0	0	0	0
Carbon Monoxide	0	0	0	0	0	0	0	0	0	0	0	0	0	0	0
Carbon Dioxide	0	0	0	0	0	0	0	0	0	0	0	0	0	0	0
Methane	0	0	0	0	0	0	0	0	0	0	0	0	0	0	0
Acetylene	0	0	0	0	0	0	0	0	0	0	0	0	0	0	0
Ethylene	0	0	0	0	0	0	0	0	0	0	0	0	0	0	0
Ethane	0	0	0	0	0	0	0	0	0	0	0	0	0	0	0
Propane	0	0	0	0	0	0	0	0	0	0	0	0	0	0	0
Water	55644.2734	55644.2734	105000	105000	105000	135112.5	139648.406	139648.406	386270.031	386270.031	481.1948	481.1948	251447	251447	123479.414
Sulphur	0	0	0	0	0	0	0	0	0	0	0	0	0	0	0
Carbonyl Sulfide	0	0	0	0	0	0	0	0	0	0	0	0	0	0	0
Hydrogen Sulfide	0	0	0	0	0	0	0	0	0	0	0	0	0	0	0

Table B-1. (Continued)

Stream No.	1931	1932	1935	1936	1938	1940	1941	1942	1943	1944	1945	1946	1950	1951	1952
Ammonia	0	0	0	0	0	0	0	0	0	0	0	0	0	0	0
HydrogenChloride	0	0	0	0	0	0	0	0	0	0	0	0	0	0	0
Silicon Dioxide	0	0	0	0	0	0	0	0	0	0	0	0	0	0	0
Calcium Oxide	0	0	0	0	0	0	0	0	0	0	0	0	0	0	0
Benzene	0	0	0	0	0	0	0	0	0	0	0	0	0	0	0
Naphthalene	0	0	0	0	0	0	0	0	0	0	0	0	0	0	0
Hybrid Poplar Ch	0	0	0	0	0	0	0	0	0	0	0	0	0	0	0
Sulfur Dioxide	0	0	0	0	0	0	0	0	0	0	0	0	0	0	0
Hydrogen Cyanide	0	0	0	0	0	0	0	0	0	0	0	0	0	0	0
Nitric Oxide	0	0	0	0	0	0	0	0	0	0	0	0	0	0	0
Methanol	0	0	0	0	0	0	0	0	0	0	0	0	0	0	0
Ethanol	0	0	0	0	0	0	0	0	0	0	0	0	0	0	0
Isopropanol	0	0	0	0	0	0	0	0	0	0	0	0	0	0	0
N-Propanol	0	0	0	0	0	0	0	0	0	0	0	0	0	0	0
Isobutanol	0	0	0	0	0	0	0	0	0	0	0	0	0	0	0
N-Butanol	0	0	0	0	0	0	0	0	0	0	0	0	0	0	0
1-Pentanol	0	0	0	0	0	0	0	0	0	0	0	0	0	0	0
RDF	0	0	0	0	0	0	0	0	0	0	0	0	0	0	0

In: Using Municipal Solid Waste for Fuel   ISBN: 978-1-61209-512-7
Editor: Samantha M. Feller   © 2011 Nova Science Publishers, Inc.

Chapter 3

METHODOLOGY FOR ALLOCATING MUNICIPAL SOLID WASTE TO BIOGENIC AND NON-BIOGENIC ENERGY

Energy Information Administration

1. INTRODUCTION

Heightened interest in renewable energy has prompted the Energy Information Administration (EIA) to examine some aspects of how it classifies energy sources as renewable. EIA employs the following definition of renewable energy sources: "Energy resources that are naturally replenishing but flow-limited. They are virtually inexhaustible in duration but limited in the amount of energy that is available per unit of time. Renewable energy resources include: biomass, hydro, geothermal, solar, wind, ocean thermal, wave action, and tidal action." Note that this definition defines renewable energy according to its primary source, which contrasts with other definitions that define any recurring waste stream as renewable.[1]

One concern in defining renewable energy fuels is how municipal solid waste (MSW) should be classified. Historically, because MSW has widely been viewed as principally composed of biomass, EIA has classified all consumption at MSW combustion plants as a renewable portion of "Waste Energy."[2] However, according to EIA's definition above, MSW clearly contains non-renewable components, raising a concern that EIA has been overstating the renewable content of MSW.

EIA recognizes that definitions of renewable energy used for State and federal energy policy purposes differ widely as to whether and to what extent MSW is included. For example, some States renewable portfolio standard (RPS) programs include all or part of MSW-fueled generation as an RPS-eligible generation source, while others do not. At the federal level, the treatment of MSW as a form of renewable energy varies across programs, laws, and even across sections of a given statute. For example, the definition of renewable energy in Section 203 of the Energy Policy Act of 2005 explicitly includes MSW-derived electricity as a "renewable energy" resource eligible to satisfy the federal renewable energy purchase requirement established in that section. Yet, many other sections of the same bill do not include MSW as an eligible renewable energy source for purposes of programs that aim to develop, assess, or support renewable energy.

To address this issue, EIA investigated whether sufficient information exists to reasonably divide MSW into its biogenic and non-biogenic portions. As a result, EIA has concluded that sufficient information does exist to reasonably estimate the split of energy produced from biogenic and non-biogenic components of MSW.

As a source of policy-neutral energy data, it is important for EIA to apply a consistent approach to defining renewable energy in its standard data reports. EIA will now include MSW in renewable energy only to the extent that the energy content of the MSW source stream is biogenic. This approach is more consistent with the definition of renewable energy used by EIA than alternatives that would either include or exclude all MSW from renewable energy. EIA's treatment of MSW's contribution to an aggregate measure of renewable energy is not intended to infringe on the prerogative of policymakers at the federal and/or State level to adopt one or more definitions of renewable energy that use a different approach to classifying MSW for their distinct policy purposes. Rather, our aim is to present energy data in a clear and consistent way, with enough detail that others can develop data consistent with whatever definition suits their objective.

MSW is primarily composed of residential solid waste but also includes some types of non-hazardous commercial, institutional and industrial wastes. MSW can be problematic to discard because of its large volume: one commonly adopted solution is to combust the MSW, which both decreases the volume of material and creates energy that can be recovered in the form of heat or steam. Because some materials have higher heat content than others, the amount of energy that can be produced by combusting MSW is a function of the composition of the waste stream.[3] For example, certain types of plastics have more than three times the heat content of yard trimmings or organic textiles. In general, combustible non-biogenic materials are characterized by higher heat contents per unit weight than combustible biogenic materials. Thus, the ratio of biogenic to non-biogenic material volumes can have a considerable effect on the heat content of the waste stream.

2. SUMMARY OF ACTIVITIES

Using data from EIA, the Environmental Protection Agency (EPA), and fuel-specific Btu values, EIA determined two interrelated trends in the composition of the MSW stream. First, the heat content (per unit weight) of the waste stream has been steadily increasing over time. Second, the shares of energy contributed to the waste stream by biogenic and non-biogenic components have been changing over time. In 1989, biogenic materials contributed two-thirds of the heat content of the waste stream. By 2005, that number had dropped to 56 percent (see Table 1 and Figure 1). This change can likely be attributed to the changing composition of the MSW stream, as increasingly more plastics are being discarded at the same time that decreasing amounts of paper and paper products are entering the waste stream.

Based on this historical information, EIA has developed estimates of biogenic and nonbiogenic MSW energy consumption (Table 2) to aid EIA in predicting the change in MSW composition over time. In keeping with the above trends, EIA estimates that nonbiogenic MSW energy consumption grew slightly faster than total MSW energy consumption over the period from 1989-2005.

EIA is applying these results beginning with data for 2006, reporting only the biogenic portion of MSW as renewable energy for purposes of the National Energy Information System. The non-biogenic portion will be reported as a component of Other Non- Renewable Waste.[4] That is, MSW consumption data will be divided into renewable and non-renewable energy based on the estimated heat content of the biogenic and nonbiogenic portions of MSW (Table 1). MSW data will be revised back to 2001 in the *Electric Power Monthly* (*EPM*) (Table 2). EIA is publishing MSW generation and consumption split into its biogenic (renewable) and non-biogenic (non-renewable) portions in the March 2007 publications of the *Monthly Energy*

Review and the *EPM*, which publish December 2006 preliminary data for the first time, and revises 2001 through November 2006 data.

The remainder of this article describes the history of MSW and the methodology EIA will use to estimate the portions of energy derived from biogenic and non-biogenic MSW.

Table 1. Municipal Solid Waste (MSW) Heat Content and Biogenic/Non-Biogenic Shares, 1989-2005

Year	Heat Content (Million Btu/Ton)	Shares of Total MSW Energy	
		Biogenic	Non-Biogenic
1989	10.08	0.67	0.33
1990	10.21	0.66	0.34
1991	10.40	0.65	0.35
1992	10.61	0.64	0.36
1993	10.94	0.64	0.36
1994	11.15	0.63	0.37
1995	11.11	0.62	0.38
1996	10.94	0.61	0.39
1997	11.17	0.60	0.40
1998	11.06	0.60	0.40
1999	10.95	0.60	0.40
2000	11.33	0.58	0.42
2001	11.21	0.57	0.43
2002	11.19	0.56	0.44
2003	11.17	0.55	0.45
2004	11.45	0.55	0.45
2005	11.73	0.56	0.44

Note: Years in bold are EPA data collection years. Non-bolded years have been linearly interpolated at the materials group level between immediately surrounding bolded years.

Sources: Heat Content (Million Btu/ton) is derived from Environmental Protection Agency, *Municipal Solid Waste in the United States: 2005 Facts and Figures*, Table 4. http://www.epa.gov/msw/msw99.htm Biogenic and non-biogenic percentages are EIA estimates.

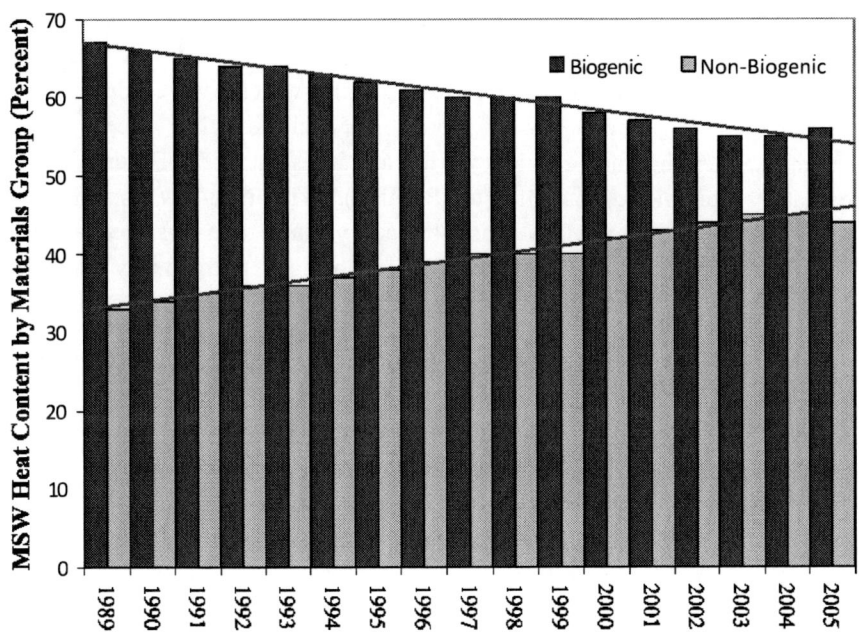

Source: Table 1 and Environmental Protection Agency, *Municipal Solid Waste in the United States: 2005 Facts and Figures*. http://www.epa.gov/msw/msw99.htm.

Figure 1. Trends in Municipal Solid Waste (MSW) Composition.

Table 2. Municipal Solid Waste (MSW) Consumption: Biogenic (Renewable) and Non-Biogenic Energy (Non-Renewable) (Trillion Btu).

	2001	2002	2003	2004	2005
Total	289	325	293	299	299
Biogenic (Renewable)	165	182	161	164	167
Non-Biogenic (Non-Renewable)	124	143	132	135	132

Sources: Total MSW consumption: Form EIA-906, "Power Plant Report" and Form EIA-920, "Combined Heat and Power Plant." Biogenic (Renewable) and non-biogenic (non- renewable) shares: EIA estimate.

3. HISTORY

Although the first facility that combusted MSW for energy came on line in New York City in 1898, the industry did not experience rapid growth until

1978 with the enactment of the Public Utility Regulatory Policy Act (PURPA).[5] This legislation made it mandatory for utilities to purchase electricity from qualifying facilities (QFs), which were defined as "cogeneration or small power production facilities that meet certain ownership, operating, and efficiency criteria established by the Federal Energy Regulatory Commission pursuant to (PURPA)." This new law improved the economics of the many MSW waste-to-energy plants that qualified as QFs. PURPA mandated the price paid for electricity be equal to the utility's avoided cost of energy and capacity, and this resulted in MSW QFs receiving a higher price for their power than they might otherwise have received.[6] MSW plants also benefited from the increased cost of landfilling due to increases in "tipping fees" (the cost to dump waste at a landfill), making disposing of MSW at a waste-to-energy plant less expensive than at a landfill in many cases.

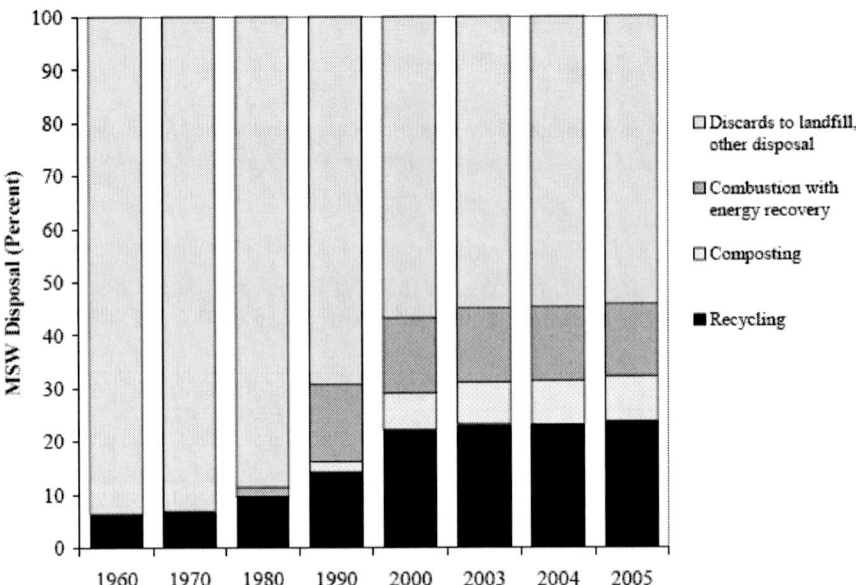

Source: Table 3 and Environmental Protection Agency, *Municipal Solid Waste in the United States: 2005 Facts and Figures*. http://www.epa.gov/msw/msw99.htm.

Figure 2. Municipal Solid Waste Disposal (MSW).

MSW waste-to-energy plants have high capital costs, and in order to make these plants financially viable, project financers required the plant to obtain a reliable stream of low- cost fuel. Usually, a plant would enter into a "flow contract" in which a municipality delivered its waste stream to a specific plant. Thus, certain facilities held a *de facto* monopoly over a certain locality's MSW. In some cases, these contracts were seen as restricting interstate commerce in municipal wastes, and in 1994 the U.S. Supreme Court upheld a challenge to flow control, finding that it violated the interstate commerce clause of the Constitution. This ruling partially or fully voided many flow supply contracts and created an added constraint on the waste-to-energy industry. Subsequent to this ruling, few plants have been able to come on line.

EIA tracks all electricity-generating plants with a capacity greater than 1 megawatt on survey forms EIA-906 "Power Plant Report" and EIA-920 "Combined Heat and Power Plant Report", including information on all MSW combustion facilities that meet this minimum capacity requirement. Historically, MSW and landfill gas (LFG) have been reported as the same fuel code on these surveys, but starting in 2001, MSW and LFG were reported as separate fuels. Therefore EIA has data that depict total MSW consumption from the year 2001 onward.

EPA maintains historical data on MSW (Table 3 below).

Table 3. Municipal Solid Waste (MSW) Disposal (Percent by Weight)

Year	1960	1970	1980	1990	2000	2003	2004	2005
Total materials recovery	6.4	6.6	9.6	16.2	29.1	31.1	31.4	32.1
Recycling	6.4	6.6	9.6	14.2	22.2	23.2	23.1	23.8
Composting [a]	s	s	s	2	6.9	7.9	8.3	8.4
Combustion with energy recovery [b]	0	0.3	1.8	14.5	14.2	14	13.8	13.6
Discards to landfill, other disposal [c]	93.6	93.1	88.6	69.3	56.7	54.9	54.8	54.3

[a] Composting of yard trimmings, food scraps, and other MSW organic material. Does not include backyard composting.

[b] Includes combustion of MSW in mass burn, modular, and refuse-derived fuel plants.

[c] Includes all other MSW that is not recovered for recycling, composting or combustion (with energy recovery). These discards are generally disposed of in landfills.

Note: Details may not add to totals due to rounding.
Source: Environmental Protection Agency, *Municipal Solid Waste in the United States: 2005 Facts and Figures*. http://www.epa.gov/msw/msw99.htm s=value less than 0.01 percent of total.

4. DATA SOURCES

Splitting MSW energy inputs into biogenic and non-biogenic components required three categories of data. First, the total amount of energy input to MSW combustion plants was obtained from EIA Forms EIA-906 and EIA-920 (Table 2). Second, data concerning the composition of the MSW stream were obtained from EPA's publication *Municipal Solid Waste in the United States*, which has been published yearly or bi-annually since 1995 (see Appendix, Table A1 for data from 1989-2005). Third, the energy content applicable to each of the combustible materials in the MSW plant input stream was obtained from the best sources EIA could identify. For most biomass fuels, EIA used Btu values published in Table B6 of the *Renewable Energy Annual* (Average Heat Content of Selected Biomass Fuels), http://www.eia.doe.gov/cneaf/solar.renewables/page/rea_data/tableb6.html. The heat content values for the remaining biomass fuels and all non-renewable fuels were obtained from various non-EIA sources. Table 4 lists the Btu content used for each component of the waste stream and the sources for that data. In some cases, the Btu value calculated was an average of several sources.

5. METHODOLOGY USED TO ESTIMATE BIOGENIC AND NON-BIOGENIC-SOURCED MSW

Beginning in 1989 and for each year that EPA data exist, the potential quantities of combustible MSW discards (which include all MSW material available for combustion with energy recovery, discards to landfill, and other disposal) shown in Table A1 were multiplied by their respective Btu contents (Table 4). These EPA-based categories of MSW were then classified into biogenic and non-biogenic groupings (Table 5). From this, EIA was able to calculate how much of the energy potentially consumed from MSW should be attributed to renewables (biogenic) and how much should be put in the category of "other, non-renewable (non-biogenic) fuels."

Table 4. Typical Heat Content of Materials in Municipal Solid Waste (MSW) (Million Btu Per Ton)

Materials	Million Btu per ton
Plastics	
Polyethylene terephthalate [c, e] (PET)	20.5
High density polyethylene [e] (HDPE)	38.0
Polyvinyl chloride [c] (PVC)	16.5
Low density polyethylene/ Linear low density polyethylene e	
(LDPE/LLDPE)	24.1
Polypropylene [c] (PP)	38.0
Polystyrene [c] (PS)	35.6
Other [e]	20.5
Rubber [b]	26.9
Leather [d]	14.4
Textiles [c]	13.8
Wood [b]	10.0
Food [a, c]	5.2
Yard trimmings [b]	6.0
Newspaper [c]	16.0
Corrugated Cardboard [c, d]	16.5
Mixed paper [e]	6.7

a Includes recovery of other MSW organics for composting.

b Energy Information Administration, Renewable Energy Annual 2004, "Average Heat Content of Selected Biomass Fuels," (Washington, DC, 2005).

c Penn State Agricultural College Agricultural and Biological Engineering and Council for Solid Waste Solutions, Garth, J. and Kowal, P. Resource Recovery, Turning Waste into Energy, University Park, PA, 1993.

d Bahillo, A. et al. Journal of Energy Resources Technology, "NOx and N2O Emissions During Fluidized Bed Combustion of Leather Wastes," Volume 128, Issue 2, June 2006. pp. 99-103.

e Utah State University Recycling Center Frequently Asked Questions. http://www.usu.edu/recycle/faq.htm.

Table 5. Municipal Solid Waste (MSW) Material Categories in Biogenic and Non-Biogenic Groups

Biogenic	Non-Biogenic
Newsprint	Plastics
Paper	PET
Containers & packaging	HDPE
Textiles	PVC
Yard trimmings	LDPE/LLDPE
Food wastes	PP
Wood	PS
Other biogenic	Other plastics
Leather	Rubber
	Other non-biogenic

Note: For explanation of plastics abbreviations, see Appendix C.

Specifically, the procedure to calculate the division is as follows: (For calculations with 2005 as an example, see Appendix B)

1. Obtain the total weight of MSW discarded in a data year (Table A1).
2. For materials that are not combusted, set Btu values to zero, (e.g., glass and metals).[7]
3. Separate the remaining MSW weights into biogenic and non-biogenic fuel groups (Table 5). These two groups represent the weight of biogenic and non-biogenic materials that are theoretically available to MSW plants for combustion.
4. Multiply the Btu per ton factors from Table 4 by the total number of tons of each materials group estimated to be available for combustion. This provides both a weighted average estimate of the total Btu

available to MSW plants for each EPA fuel group and a total weighted Btu per ton value for an average ton of MSW.

5. Calculate the biogenic and non-biogenic percentages of the estimated waste stream input to MSW combustion facilities.
6. Multiply these percentages by the MSW consumption data (in Btu) to estimate the total Btus available from biogenic versus non-biogenic sources (Table 2).
7. Repeat for all data years, and interpolate linearly between years to obtain the final estimates of shares of biogenic and non-biogenic MSW energy consumption (Table 1).

5.1. Caveats and Assumptions

1. The data provided by EPA are available only at the national level. Therefore, the assumed rates of recycling or MSW generation may not hold true for all States and regions.
2. The composite MSW HHV by year (Table 1) is not representative of the MSW HHV reported by EIA respondents. This is because respondent data did not include MSW composition information, which was necessary to divide MSW into biogenic and non-biogenic shares. For the purposes of determining a split, the small difference (between EIA respondent MSW HHV and calculated MSW HHV) most likely does not affect the results.
3. Discards available for combustion represent the average mixture of MSW available on a national level, which may or may not be representative of the mixture of MSW obtained by any particular MSW combustion facility.
4. Btu values were obtained from best available sources and may have been developed using different procedures.
5. All fuels have a range of Btu values from lower heating value (LHV) to higher heating value (HHV).[8] If the source did not specify which of these the value represented, it was assumed to represent HHV.
6. The EPA data divide categories of materials into sub-groups (for example, plastics are divided into seven sub-groups), each of which has a specific Btu content. However, in some cases, it is not clear what is included in the "other" category within a fuel group. Because of this, "other plastics" was assigned a heat content equal to the value of the average of all plastics groups not in the "other" category.

7. Unassigned "Other" is divided evenly between the renewable and non-renewable groups.

8. MSW input materials were classified as biogenic or non-biogenic according to their predominant composition.
   a. All rubber is assumed to be tires with the metals, fiber, and other material removed. Therefore, the rubber is assumed to be synthetic, and is treated as a non- biogenic material.[9]
   b. Textiles are all assumed to be composed of biogenic materials and are thus treated as biogenic.

9. Depending on plant design, some MSW plants may need to remove glass and metals prior to combustion. Because of their poor combustion characteristics, these materials do not produce significant heat when combusted, and may in fact reduce the net heat output of a plant. In order to allow for uniform analysis of waste stream heat contents, the Btu values of glass and metal were set at zero.

10. For the years EPA did not collect data, values for Btu content and the biogenic/non-biogenic split were interpolated linearly between the nearest prior and subsequent survey years.

6. RESULTS

As a result of this analysis, EIA has decided to split MSW into biogenic and non-biogenic components for its future data releases. This biogenic component of MSW will also be reflected in restatements of prior data back to 2001 when EIA began to collect MSW as a separate fuel.

In standard energy data reports, EIA will now include MSW in renewable energy only to the extent that the energy content of the MSW source stream is biogenic. This approach is more consistent with the definition of renewable energy used by EIA than alternatives that would either include or exclude all MSW from renewable energy. EIA's treatment of MSW's contribution to an aggregate measure of renewable energy is not intended to infringe on the prerogative of policymakers at the federal and/or State level to adopt one or more definitions of renewable energy that use a different approach to classifying MSW for their distinct policy purposes. Rather, our aim is to present energy data in a clear and consistent way, with enough detail that others can develop data consistent with whatever definition suits their objective.

As discussed above, the calculations performed revealed two trends.

1. The Btu content per ton of MSW is steadily increasing over time (Table 1).
2. The percentage of MSW's total energy content that comes from biogenic resources is decreasing, while the percentage of energy content from non-biogenic resources is increasing (Figure 1).

These trends are clearly interrelated and are most likely attributable to two factors. First, the percentage of paper and paperboard recycled has increased from 40 percent by weight in 1996 (the first year for which complete data are available) to 50 percent in 2005, while the amount combusted rose by only 3 percent (from 81.5 million tons to 84 million tons). Second, the amount of plastic combusted has increased dramatically, from 19 million tons in 1996 to 28.9 million tons in 2005, while the recycling rate grew only half of 1 percent (from 5.2 percent to 5.7 percent). These two factors have combined to decrease the amount of biogenic material in the waste stream while increasing the amount of nonbiogenic material, mainly plastics, which have one of the highest Btu contents of any fuel group in MSW.

The historical trends observed in biogenic/non-biogenic MSW composition and energy content may change as a result of EPA's national policy designed to increase the recycling rates of all materials in MSW. However, it is difficult to predict if this policy will have differing effects on the biogenic and non-biogenic components of MSW and therefore EIA cannot forecast what effect this policy will have on the heat content of the waste stream. Until the effect of EPA's policy becomes clear, EIA expects that the energy content of MSW will continue to grow over time. EIA will continue to update its estimates of the biogenic and non-biogenic portions of MSW as revised EPA data become available. For purposes of standard EIA data reports, only the biogenic portion of MSW will be included in aggregate measures of renewable energy.

APPENDIX A. DATA BY MATERIALS CATEGORY

Table A1. Municipal Solid Waste (MSW) Weights by Category, 1989-2005 (Million Tons)

Discards	1989	1990	1991	1992	1993	1994	1995	1996	1997	1998	1999	2000	2001	2002	2003	2004	2005
Textiles	4.87	5.15	5.30	5.48	5.64	5.80	6.50	6.70	7.10	7.50	7.90	8.10	8.40	8.74	9.08	9.24	9.40
Yard Trimmings	30.47	30.80	29.00	27.20	25.40	23.60	20.80	17.20	16.20	15.65	15.10	11.90	12.20	12.35	12.50	12.37	12.23
Food Wastes	20.02	20.80	19.00	17.20	15.40	13.60	13.40	21.40	21.30	22.95	24.60	25.20	25.50	26.18	26.85	27.68	28.50
Wood	11.57	12.08	12.36	12.64	12.92	13.20	13.50	10.30	11.00	11.30	11.60	12.20	11.90	12.11	12.32	12.46	12.59
Other	5.30	5.40	5.53	5.65	5.78	5.90	6.00	6.10	6.30	6.40	6.50	8.60	6.80	6.88	6.96	7.63	8.30
Glass	10.86	10.47	10.40	10.34	10.34	10.20	9.70	9.20	9.10	9.40	9.70	9.90	10.20	10.18	10.15	10.09	10.02
Rubber	4.13	4.24	4.34	4.43	4.53	4.63	4.73	4.90	5.01	5.09	5.17	4.77	4.49	4.66	4.83	4.85	4.88
Leather	1.15	1.18	1.21	1.24	1.26	1.29	0.77	0.80	0.80	0.82	0.83	0.83	0.88	0.89	0.90	0.88	0.86
Metals	12.75	12.58	11.91	11.24	11.24	9.90	9.60	9.70	9.90	10.70	11.50	11.66	11.80	11.88	11.96	11.85	11.74
Paper & Paperboard																	
Newsprint	7.27	7.40	7.40	7.41	7.41	7.41	6.17	5.64	6.12	6.03	5.94	6.28	4.85	3.54	2.23	4.13	6.03
Paper	23.99	24.41	24.42	24.42	24.43	24.44	24.58	23.78	24.49	25.11	25.72	23.78	23.00	23.48	23.96	20.20	16.44
Containers & Packaging	20.33	20.69	20.70	20.71	20.71	20.72	18.17	17.90	18.31	17.91	17.51	17.31	17.27	17.11	16.94	18.24	19.53
Plastics																	
PET	1.17	1.24	1.28	1.32	1.36	1.40	1.33	1.35	1.54	1.69	1.84	2.06	2.11	2.29	2.46	2.58	2.70
HDPE	2.76	2.93	3.03	3.12	3.22	3.31	3.15	3.17	4.21	4.37	4.52	4.41	4.49	4.58	4.67	5.11	5.55
PVC	1.31	1.40	1.44	1.49	1.53	1.58	1.50	1.23	1.32	1.35	1.37	1.39	1.42	1.45	1.47	1.51	1.55

Table A1. (Continued)

Discards	1989	1990	1991	1992	1993	1994	1995	1996	1997	1998	1999	2000	2001	2002	2003	2004	2005
LDPE/LLDPE	4.42	4.70	4.85	5.00	5.15	5.29	5.04	4.90	5.28	5.24	5.20	5.59	5.73	5.90	6.06	6.07	6.08
PP	2.44	2.59	2.67	2.76	2.84	2.92	2.78	2.45	2.67	2.67	2.67	3.34	3.45	3.53	3.60	3.69	3.77
PS	1.95	2.08	2.14	2.21	2.28	2.34	2.23	1.96	2.09	2.12	2.15	2.28	2.29	2.28	2.27	2.36	2.44
Other	1.72	1.83	1.88	1.94	2.00	2.06	1.96	3.10	3.24	3.33	3.41	4.30	4.50	4.62	4.73	4.95	5.17
Total	168.49	172	168.9	165.8	163.4	159.6	151.9	151.8	156	159.6	163.2	163.9	161.3	162.6	163.9	165.9	167.8

Notes: Totals may not equal the sum of all components due to independent rounding. Discards includes all MSW material available for combustion with energy recovery, discards to landfill, and other disposal. For explanation of plastics abbreviations, see Appendix C.

Source: Environmental Protection Agency, Municipal Solid Waste in the United States, 2005. http://www.epa.gov/msw/msw99.htm.

Table A2. Paper and Paperboard Products Weights in Municipal Solid Waste (MSW) by Category, 2005 (Thousand Tons)

Material	Generation	Recovery	Discards
Newsprint	9,790	8,730	1,060
Groundwood Inserts	2,260	1,980	280
Books	1,120	260	860
Magazines	2,520	970	1,550
Office Papers a	6,580	4,120	2,460
Telephone Directories	660	120	540
Standard Mail b	5,830	2,090	3,740
Other Commercial Printing	7,340	760	6,580
Tissue Paper and Towels	3,430	s	3,430
Paper Plates and Cups	970	s	970
Other Non-Packaging Paper c	4,410	s	4,410
Corrugated Boxes	30,930	22,100	8,830
Milk Cartons	420	s	420
Folding Cartons	4,970	590	4,380
Other Paperboard Packaging	150	s	150
Bags and Sacks	1,190	250	940
Other Paper Packaging	1,370	s	1,370
Total Paper and Paperboard	83,940	41,970	41,970

[a] High-grade papers such as copy paper and printer paper.

[b] Formerly called Third Class Mail by the U.S. Postal Service.

[c] Includes tissue in disposable diapers, paper in games and novelties, cards, etc. s=value less than 5,000 tons.

Notes: Details may not add to totals due to rounding. Discards includes all MSW material available for combustion with energy recovery, discards to landfill, and other disposal.

Source: Environmental Protection Agency, Municipal Solid Waste in the United States, 2005. http://www.epa.gov/msw/msw99.htm.

Table A3. Plastic Products Weights in Municipal Solid Waste (MSW) by Category, 2005 (Thousand Tons)

Resin	Generation	Recovery	Discards
PET	2,860	540	2,320
HDPE	5,890	520	5,370
PVC	1,640	NA	1,640
LDPE/LLDPE	6,450	190	6,260
PP	4,000	10	3,990
PS	2,590	NA	2,590
Other resins	5,480	390	5,090
Total Plastics in MSW	28,910	1,650	27,260

Notes: Some detail of recovery by resin omitted due to lack of data. This table understates the recovery of plastics due to the dispersed nature of plastics recycling activities. Discards includes all MSW material available for combustion with energy recovery, discards to landfill, and other disposal. For explanation of plastics abbreviations see Appendix C.

Source: Environmental Protection Agency, Municipal Solid Waste in the United States, 2005. http://www.epa.gov/msw/msw99.htm.

Table A4. Rubber and Leather Products Weights in Municipal Solid Waste (MSW) by Category, 2005 (Thousand Tons)

Material	Generation	Recovery	Discards
Rubber in Tires [a]	2,760	960	1,800
Other Durables [b]	2,920	s	2,920
Clothing and Footwear	700	s	700
Other Non-durables	290	s	290
Containers and Packaging	30	s	30
Total Rubber & Leather	6,700	960	5,740

[a] Automobile and truck tires. Does not include other material in tires.

[b] Includes carpets, rugs, and other miscellaneous durables. s=value less than 5,000 tons.

Notes: Details may not add to totals due to independent rounding. Discards include all MSW material available for combustion with energy recovery, discards to landfill, and other disposal.

Source: Environmental Protection Agency, *Municipal Solid Waste in the United States*, 2005. http://www.epa.gov/msw/msw99.htm.

APPENDIX B. CALCULATIONS FOR 2005 AS AN EXAMPLE

Table B1. Calculations to Obtain Average Million Btu Per Ton for Municipal Solid Waste (MSW)

Material Group	Discards (million tons)[a]	Heat Content (million Btu per ton) [b]	Heat Contributed (trillion Btu)
Newsprint	6.03	16.00	96.48
Paper	16.44	6.70	110.15
Containers &Packaging	19.53	16.50	322.25
Plastics			
PET	2.70	20.45	55.22
HDPE	5.55	19.00	105.45
PVC	1.55	16.50	25.58
LDPE/LLDPE	6.08	24.10	146.53
PP	3.77	38.00	143.26
PS	2.44	35.60	86.86
Other	5.17	20.50	105.99
Rubber	4.88	26.86	131.08
Leather	0.86	14.40	12.38
Textiles	9.40	13.80	129.72
Yard Trimmings	12.23	6.00	73.38
Food Wastes	28.50	5.20	148.20
Wood	12.59	9.96	125.40
Other	8.30	18.10	150.23
Glass	10.02	0.00	0.00
Metals	11.74	0.00	0.00
Total	167.78	11.73	1968.14
Total Btu/Total Tons →	1968.14 ÷167.78 = 11.73 Million Btu/Ton of MSW		

[a] Table A1.

[b] Table 4.

Notes: Discards includes all MSW material available for combustion with energy recovery, discards to landfill, and other disposal. For explanation of plastics abbreviations, see Appendix C.

Source: Environmental Protection Agency, Municipal Solid Waste in the United States, 2005. http://www.epa.gov/msw/msw99.htm.

Table B2. Biogenic and Non-Biogenic Contributions to Total Million Btu/Million Ton of Municipal Solid Waste (MSW)

Biogenic	Share	Non-Biogenic	Share
Newsprint	0.05	Plastics	
Paper	0.06	PET	0.03
Containers &Packaging	0.16	HDPE	0.05
Leather	0.01	PVC	0.01
Textiles	0.07	LDPE/LLDPE	0.07
Yard Trimmings	0.04	PP	0.07
Food Wastes	0.08	PS	0.04
Wood	0.06	Other plastics	0.05
Other Biogenic	0.04	Rubber	0.07
		Other Non-Biogenic	0.04
Total	0.56	Total	0.44

Note: For explanation of plastics abbreviations, see Appendix C.
Source: Biogenic and non-biogenic estimation methodology documented in this report.

APPENDIX C. ABBREVIATIONS

Btu	–	British Thermal Units
EIA	–	Energy Information Administration
EPA	–	Environmental Protection Agency
HDPE	–	High density polyethylene
HHV	–	Higher heating value
LDPE/LLDPE	–	Low density polyethylene/ Linear low density polyethylene
LFG	–	Landfill gas
LHV	–	Lower heating value
MSW	–	Municipal solid waste
PET	–	Polyethylene terephthalate
PVC	–	Polyvinyl chloride
PP	–	Polypropylene PS - Polystyrene
RDF	–	Refuse derived fuel

REFERENCES

[1]   Bahillo, A. et al. (2006). *Journal of Energy Resources Technology,* "NOx and N2O Emissions During Fluidized Bed Combustion of Leather Wastes." Volume 128, Issue 2, June, 99-103.

[2]   Energy Information Administration. *Annual Energy Review 2005.* "Renewable Energy Consumption by Source, Selected Years, 1949-2005." Washington, DC, 2006.

[3]   Energy Information Administration. *Renewable Energy Annual 2004.* "Average Heat Content of Selected Biomass Fuels." Washington, DC, 2005.

[4]   Environmental Protection Agency. *Municipal Solid Waste in the United States: 2005 Facts and Figures.* Also, years 1995, 1996, 1997, 1998, 1999, 2000, 2001, 2003. Published at *http://www.epa.gov/msw/ msw99.htm* Accessed December 2006.

[5]   Penn State Agricultural College Agricultural and Biological Engineering and Council for Solid Waste Solutions. Garthe, J. and Kowal, P. *Resource Recovery, Turning Waste into Energy,* University Park, PA, 1993.

[6]   Rubber Manufacturer's Association. *Scrap Tire Characteristics.* http://www.rma.org/scrap_tires/scrap_tire_markets/scrap_tire_characteri stics/. Accessed February 2007.

[7]   Utah State University Recycling Center Frequently Asked Questions. Published *at* http://www.usu.edu/recycle/faq.htm Accessed December 2006.

End Notes

[1] Webster's Online Dictionary, for example, defines renewable as "Capable of being replaced by ecological cycles or sound management practices."

[2] Energy Information Administration, *Annual Energy Review,* Table 10.1, Renewable Energy Consumption by Source, http://www.eia.doe.gov/emeu/aer/pdf/pages/sec10_3.pdf).

[3] Heat content is measured in British thermal units (Btu) by weight

[4] This renewable component of MSW will also be reflected in restatements of prior data back to 2001.

[5] Public Law 95-617, Public Utility Regulatory Policies Act of 1978.

[6] The incremental cost to the electric utility of alternative electric energy which the utility would have generated or purchased from another source

[7] This does not mean that recycling such materials has no net effect on energy use. Rather, it reflects the fact that waste-to-energy facilities do not derive any energy from these sources.

[8] The heat content rates (i.e., thermal conversion factors) can either represent the gross (HHV) energy content of the fuels, or the net (LHV) energy content. HHV rates are applied in all Btu calculations for EIA's *Annual Energy Review* and are commonly used in energy calculations in the United States; LHV rates are typically used in European energy calculations. The difference between the two rates is the amount of energy that is consumed to vaporize water created during the combustion process. Generally, the difference rages from 2 percent to 10 percent, depending on the specific fuel and its hydrogen content. Some fuels, such as unseasoned wood, can be more than 40 percent different in their gross and net heat contents.

[9] Rubber Manufacturer's Association, Scrap Tire Characteristics, http://www.rma.org/scrap_tires/scrap_tire_markets/scrap_tire_characteristics, accessed February 2007.

CHAPTER SOURCES

The following chapters have been previously published:

Chapter 1 – This is an edited reformatted and augmented version of a United States Department of Energy publication, report PNNL-18144, dated December 2008.

Chapter 2 – This is an edited reformatted and augmented version of a United States Department of Energy publication, report PNNL-18482, dated April 2009.

Chapter 3 – This is an edited reformatted and augmented version of a United States Department of Energy publication, report *Methodology for Allocation Municipal Solid Waste to Biogenic and Non-Biogenic Energy*, dated May 2007.

INDEX